# Monographs in Virology

Vol. 9

Editor: J. L. Melnick, Houston, Tex.

S. Karger · Basel · München · Paris · London · New York · Sydney

# Early Interaction between Animal Viruses and Cells

Karl Lonberg-Holm
Central Research Department, Du Pont Experimental Station,
Wilmington, Del.

Lennart Philipson
Department of Microbiology, The Wallenberg Laboratory,
Uppsala University, Uppsala

8 figures and 4 tables

1974

S. Karger · Basel · München · Paris · London · New York · Sydney

# Monographs in Virology

Vol. 5
WILDY, P. (Birmingham):
Classification and Nomenclature of Viruses
VIII+81 p., 1971
ISBN 3-8055-1196-5

Vol. 6
TINSLEY, T. W. and HARRAP, K. A. (Oxford):
Moving Frontiers in Invertebrate Virology
VIII+66 p., 23 fig., 3 tab., 1972
ISBN 3-8055-1464-6

Vol. 7
BECKER, Y. (Jerusalem):
The Agent of Trachoma
XIV+99 p., 12 fig., 5 tab., 1974
ISBN 3-8055-1657-6

Vol. 8
SULKIN, S. E. and ALLEN, RAE (Dallas, Tex.):
Virus Infections in Bats
VIII+103 p., 1 fig., 14 tab., 1974
ISBN 3-8055-1696-7

S. Karger · Basel · München · Paris · London · New York · Sydney
Arnold-Böcklin-Strasse 25, CH-4011 Basel (Switzerland)

---


Printed in Switzerland by Berner Zeitung AG, Langnau
ISBN 3-8055-1764-5

# Contents

# Editor's Preface

Over the past decade a vast amount of progress has been made in understanding viral replication cycles, but during this period there have been few detailed reviews of the *early* interactions between animal viruses and cells. In this Monograph KARL LONBERG-HOLM and LENNART PHILIPSON, active investigators who themselves have contributed substantially to the advancement of knowledge in this area, deal comprehensively with the early events in interaction of virions and cells, placing in context a monumental total of 756 pertinent references from the literature. Their review, a most valuable one which represents a happy combination of expertise and readability, covers the present state of knowledge about those events which begin with attachment of the virus at the plasma membrane and proceed through host cell macromolecular synthesis during early interaction – up to the point at which virus-specific macromolecular synthesis is initiated. As requisite background, the authors include a survey of present understanding of the virion structural proteins that participate in these early interactions and of the structure of the plasma membrane. In reviewing and analyzing the biochemical and morphological evidence that is available concerning the nature of the early steps in infection, they provide sophisticated generalizations where the evidence permits. They also offer helpful commentary on possible approaches to some of the important areas of information that still remain to be clarified.

JOSEPH L. MELNICK

# Acknowledgments

The authors thank RICHARD L. CROWELL and RICHARD W. COMPANS for contributing helpful comments and corrections. We also thank BERIT NORDENSVED, D. JANE MEADOWS, ELAINE K. DANELUTTI and FREDRICKA LONBERG-HOLM for diligent aid in the organization and preparation of the manuscript and literature references.

# I. Introduction

## *A. Scope*

This review concerns the interaction of virions and cells starting with attachment and up to but not including the initiation of virus-specific macromolecular synthesis. An understanding of this process will require detailed analysis of the virion structures which recognize cellular receptors, which participate in penetration, and possibly also which influence the distribution of subviral particles within the cell. Secondly, a detailed analysis of the plasma membrane and of pertinent intracellular structures will also be needed. The structural evidence, and also our present understanding of the early events of infection, are lacunate but a review of what is known may at least show the way for future work.

During the last two decades there have been rapid advances in the accurate assay of the biological activities of virions, as well as an increased understanding of the structural design and protein composition of several animal viruses. In addition, chemical tracer techniques have been widely used to identify virions and subviral particles within infected cells. A major obstacle, however, in describing the true pathway of infection is the large number of physical particles required for each infectious unit. The ratio of physical to infectious particles ranges from 10 to 1,000 for most animal viruses and this makes it difficult to distinguish between events which lie in the infectious pathway and those which lie in an abortive avenue, irrespective of whether electron microscopic or biochemical techniques are employed to detect parental virus. Also, the early events are often asynchronous and it is exceptional when a particle can be identified as an intermediate in the infective sequence on the basis of kinetic evidence alone. These shortcomings can be attacked in two ways. First, specific inhibitors may be found which block infectious center formation and also one stage in early interaction; antibodies have also been used to inhibit early steps in infection. Secondly, either virus or cell mutants, defective in a step of early infection, would constitute powerful instruments in this study. Although there has recently been considerable development of temperature-sensitive viral mutants, very few, if any, are available for a detailed study of early interaction. Cellular mutants have been described in which early steps in interaction are ineffective, but these have not yet been employed in detailed analysis of virus-cell interaction.

The early events of virus infection have been reviewed frequently during the last decade [138, 318, 523], but comparatively infrequently during the last five years [139], despite continued research. This review will emphasize the biochemistry of early virus-cell interaction and we will attempt to make generalizations in this area. The structures involved in early virion-cell interaction (i. e. the structural proteins of the virion) and the structure of the plasma membrane will be discussed first. Then the biochemical and morphological evidence concerning the nature of the early steps in infection will be considered.

## *B. Definition of Terms*

(a) *Attachment* denotes the initial and specific union between virion and cell; this step is usually measured by the disappearance of virus from the supernatant of virus-cell mixtures. Occasionally, this step is reversible under the conditions employed for infection, i. e. cell-associated and extracellular infective virus can be shown to be in equilibrium. More often, the reversal of attachment can be demonstrated only with nonphysiological dissociating agents or in some cases after the action of a virion-associated enzyme.

(b) *Receptors* are natural components of the cell surface which bind virus. They may be discrete chemical entities or possibly mosaic patterns on the plasma membrane.

(c) *Receptor sites* are operationally defined as loci which can be occupied by virions; there may be fewer receptor sites than receptors.

(d) *Irreversible alteration* indicates a permanent alteration of the virion so that it cannot be recovered intact and infective by any known means; it is thus operationally defined. The terms *'eclipse'* or *'irreversible attachment'* have also been employed, but the latter term may by misleading since it is now evident that infectivity of nonenveloped viruses is lost as a result of structural alteration rather than as a result of the nondissociable attachment of a cellular component.

(e) *Elution* of infective virus can occur if attachment is truly reversible under physiological conditions. The term elution often refers, however, to the detachment of noninfective irreversibly altered particles and is thus an inefficient alternative to penetration. We will discuss both kinds of elution.

(f) *Penetration* refers to the sequence of events, following attachment, by which the virion reaches the site of its replication. Penetration may occur either before or after the virus has been irreversibly altered. The viral genome may pass directly through the plasma membrane because of enzymatic activity or properties inherent in the virion or subviral particle, or alternatively it may enter a cellular versicle by endocytosis or specifically by pinocytosis (a process sometimes termed 'viropexis'). It is also conceivable that a virus could be uncoated by the receptors at the outer surface of the cell and that the nucleic acid could then penetrate by whatever general mechanism exists for the cellular uptake of nucleic acids, but there is at present no evidence that this is a significant route for infection.

# II. Interacting Structures

## *A. Virus Structure*

### 1. General Aspects

Although animal viruses differ in shape, chemical composition, and size, there are common principles in their construction. Basically, a virus is a nucleic acid genome, either DNA or RNA, enclosed in a capsid of variable complexity and composed mainly of protein [408]. The nucleocapsid of the virion comprises the nucleic acid plus capsid protein. The capsid is constructed of repeating subunits, which minimizes the genetic information necessary for its synthesis [120]. The capsid is in one of two basic patterns; in one the nucleic acid is *extended* and in the other it is *condensed*. Superimposed on these two patterns are variations dictated both by the size of the genome and by the nature of the capsid polypeptides. In addition, the virus particles may consist of only a nucleocapsid, i. e. the genome is surrounded by only one layer of protein, or there may be additional protein shells. Finally, there may be one or possibly two membrane layers outside the nucleocapsid.

A prototype for extended nucleocapsids is the plant virus, tobacco mosaic virus (TMV), which is a long rod of 8 nm diameter. The nucleic acid molecule is arranged as a helix within the capsid so that there is one polypeptide for every three ribonucleotides. All the polypeptides are in equivalent positions and make the same bonds with their neighbors [12, 87, 378]. In mutant virus strains the polypeptide molecules occupy quasi-equivalent positions [85]. The orthomyxoviruses, paramyxoviruses, rhabdoviruses and possibly the RNA tumor viruses possess nucleocapsids with helical symmetry which probably are also constructed according to similar principles but with different details.

In the second type of virion the nucleic acid is condensed. It forms, either alone or in combination with one or several basic polypeptides, the central portion of a quasi-spherical nucleocapsid containing one or several shells of protein subunits. Structural analyses by X-ray diffraction indicate that the subunits are often arranged according to icosahedral symmetry [87, 120, 352] and this requires that the capsid subunits interact strongly enough with each other so that their locations are geometrically restricted.

Structural analysis by electron microscopy reveals that most capsids are composed of large morphological subunits or capsomeres [408]. The morphology of the capsomeres varies both in size and shape from one virus to another and round, prismatic, solid and hollow capsomeres have all been observed. The proteins of the virion have been resolved into polypeptides of different molecular weights by polyacrylamide-SDS-gel electrophoresis, but the relationship between crystallographic protein subunits, the capsomeres, and the relatively small polypeptides remains in many cases to be established.

The capsids of most animal viruses with cubic symmetry appear to be polyhedral, i. e. the capsomeres are located at the vertices of regular polydeltahedra. The simplest of these, the icosahedron, has 20 equilateral triangular facets. Other icosahedra have $20 \times T$ facets where T is the 'triangulation number' [87, 352, 437]. This number is given by the expression $T = Pf^2$, where f is any integer and P equals $(h^2 + hk + k^2)$; h and k are integers that have no common factor other than one. The total number of the capsomeres is related to the number of facets by the expression $10\,T + 2$. This means that not all capsomeres occupy equivalent positions as they obviously must in a simple icosahedron with $T = 1$. Thus the capsomeres at the 12 vertices of an icosahedron have 5 nearest neighbors, all others 6. There are at least 6 groups of animal viruses whose capsids have icosahedral symmetry. Table I summarizes current views on their structures; reviews on this subject are also available [86, 87, 351, 352, 437].

The relationship between the capsomeres revealed by electron microscopy and the icosahedral symmetry of the virion is based on the assumption that the subunits are comparatively simple. The virions of several plant viruses seem to contain only one distinct polypeptide [129, 252] but with the animal viruses the polypeptide composition appears to be complex. Thus simple $T = 1$ or $T = 3$ viruses like the picornaviruses [527, 578] contain four distinct polypeptides from 6,000 to 35,000 molecular weight [578, 651]. The symmetry rules, however, seem to be obeyed since all picornaviruses examined contain 60 molecules of each of these four non-equivalent polypeptides, and this was predicted to be the smallest possible number of subunits in capsids showing icosahedral symmetry by X-ray diffraction [578]. Adenoviruses with a $T = 25$ are profoundly more com-

*Table I.* Structural parameters of animal viruses with cubic symmetry[1]

| Group | Nucleic acid | Triangulation number, T | P | f | Capsomeres | Hexons | Pentons | Polypeptides | |
|---|---|---|---|---|---|---|---|---|---|
| | | | | | | | | in capsid | in core |
| Picornavirus | ssRNA | 1 or 3 | 1 or 3 | 1 | 12 or 32 | 0 or 20 | 12 | 4 | 0 |
| Reovirus | dsRNA | 9 | 1<br>3 | 3<br>2 | 92<br>122[2] | 80<br>110 | 12 | 3 | 4 |
| Adenovirus | DNA | 25 | 1 | 5 | 252 | 240 | 12 | 8–11[4] | 2 |
| Herpesvirus[3] | DNA | 16 | 1 | 4 | 162 | 150 | 12 | 4 | 2 |
| Papovavirus | DNA | 7 | 7 | 1 | 72 | 60 | 12 | 3 | 3 |

1 For a review see JOKLIK and ZWEERINK [323]. More recent data are discussed in the text.
2 Data from LUFTIG *et al.* [404].
3 Nucleocapsid without envelope.
4 Data from EVERITT *et al.* [193] and MAIZEL *et al.* [412].

plex [689] and contain at least 10 and possibly 13 polypeptides [193, 412] of which two are associated with DNA in a core structure [193]. The nucleocapsid of herpes virions comprises at least 6 polypeptides two of which are internal basic polypeptides, while the envelope contains at least 3 and perhaps as many as 8 components [328, 504, 715]. It is therefore becoming progressively clear that it will be difficult to correlate the polypeptide composition of animal viruses with the results of X-ray crystallographic and electron microscopic studies.

In some icosahedral viruses studied in detail, the vertex capsomeres with local 5-fold symmetry appear to consist of polypeptides which are different from those forming the hexon structures of the facets and edges. This has been established for an adenovirus [520] by tryptic peptide maps, amino acid composition and immunological techniques, and is probably also true for picornaviruses [312, 527]. One teleological explanation for the complexity of animal compared to plant viruses may be the fact that the former are often passively introduced into the cell by invertebrate vectors while the latter must attach to and penetrate the cell membrane on the basis of their own structural properties.

Animal viruses which have helical symmetry as well as others which do not, such as herpes- and arboviruses, are enclosed in an envelope which consists of proteins, glycoproteins, lipids, phospolipids and glycolipids. The envelopes may be derived from the outer plasma membranes of the infected cell, as in the case of myxoviruses [112, 457] and rhabdoviruses [591] whose nucleocapsids gain release from the cell by a budding process. The envelope in these cases appear to consist of a bimolecular layer of lipids with glycoprotein spikes attached in a regular manner on the outer surface [253, 679]. The nucleocapsids of RNA tumor viruses appear to be enclosed in two envelopes, one originating within the cytoplasm and the other outer envelope derived from plasma membrane when the virus 'buds' from the cell [20, 373]. The envelopes of herpesviruses, finally, appear to vary in size and composition from one to another type. They are usually acquired when the naked nucleocapsid exits from the cell nucleus [487, 604, 632] to the cytoplasm but in some cases may originate from the endoplasmic reticulum and additional material may also be acquired when the virus particle is liberated from the cell [185].

The protein composition of the enveloped viruses has been studied in great detail as exemplified with the orthomyxoviruses. Influenza virus contains 6–8 virus-specific polypeptides [112, 113, 350, 602, 628]; their relationship to the virion ultrastructural components has been reviewed

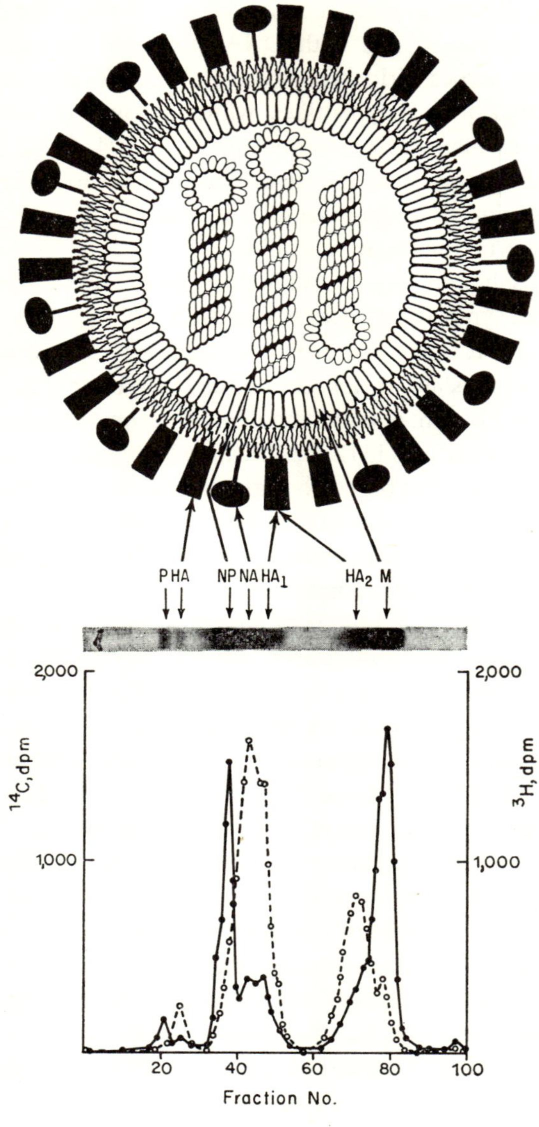


*Fig. 1.* Schematic outline of orthomyxoviruses, indicating the position and the nomenclature of the virion proteins as revealed by SDS polyacrylamide gels. The material used for gel electrophoresis was doubly labeled with $^{14}C$-amino acids (●) and $^{3}H$-glucosamine (O) indicating that the polypeptides NA, HA1, HA2 are glycosylated. The uncleaved HA polypeptide, which is a precursor to HA1 and HA2, is also glycosylated. This figure was kindly provided by Dr. R. COMPANS and will appear in

and a nomenclature for the polypeptides has been proposed [344]. Figure 1 illustrates the structure of the orthomyxoviruses.

The orthomyxovirus virion envelope, like that of the paramyxoviruses, contains a neuraminidase and lacks sialic acid [349]. There are at least 3 glycoprotein subunits of the orthomyxovirus; one is associated with the neuraminidase and two are subunits of the hemagglutinating spike [350, 603, 628, 638] and are probably cleavage products of a single glycopolypeptide precursor [344]. The intact hemagglutinin has been characterized after its separation from the envelope by detergent treatment [719, 720]. The neuraminidase likewise has been characterized after removal from the virion by detergent [380] or treatment with trypsin [165, 602]. The paramyxoviruses contain 6 polypeptides [192, 464, 465], two of which are glycopolypeptides associated with spikes on the virion envelope [95], and only one of these is responsible for both neuraminidase and attachment or hemagglutination [592, 593].

Arboviruses have a very simple structure. Their RNA is condensed with one 32,000 molecular weight protein to form a spherical nucleocapsid which does not readily show symmetry elements [210, 258, 621, 648]. This structure is surrounded by a bimolecular lipid membrane layer [253] at the outer surface of which are attached spikes composed of two species of glycoproteins [595, 596]. Poxviruses, finally, are the largest and the most complex of all animal viruses. The outer shell of these complex viruses contains at least 15 different species of polypeptides and glycopolypeptides as well as lipids and phospholipids [282, 284]. Electron microscopy reveals filaments arranged in a mulberry-like pattern [137]. Inside this shell there resides a brick-shaped core which is bounded by a distinct membrane with projections on both the outer and the inner surface [179]. In addition to the core, the poxvirus envelope encloses two 'lateral bodies' whose function is unknown [538].

The nucleic acid of animal viruses does not interact equally with all virion proteins and in many cases a distinct central 'core' is observed, composed of a limited number of proteins and the genome. The term 'core' has not been used consistently and for this reason the terminology is

---

Lenard and Compans 'The membrane structure of lipid containing viruses', BBA Reviews in Biomembranes. The gel pattern was published in [113]. The lipid bilayer between the M protein on the inside and the glycoproteins on the outside does not appear to be penetrated by proteins as outlined in section II. B for the cellular plasma membrane, but the forces holding the proteins in place and together for the virus are still unknown.

reviewed here. In the cases of vaccinia virus and reovirus, the structure usually described as a core is a regularly constructed nucleocapsid. The vaccinia core is described above and contains the DNA-protein 'chromosome' and also a DNA-dependent RNA polymerase [179, 283, 284, 447, 587]. The reovirus core consists of an icosahedral shell containing ten segments of double-stranded RNA and 4 polypeptides [404]; these cores are produced by digesting 3 outer virion polypeptides with chymotrypsin. The RNA can also be released from the core as a tightly coiled mass which may contain protein.

In other viruses with cubic symmetry, the core is defined as consisting of nucleic acid and associated proteins held within, but not including, the capsid structure. The core of adenovirus contains two noncapsid polypeptides [544] and the core of herpesvirus contains two noncapsid polypeptides in addition to polyamines which help to reduce the net negative charge [213]. The cores of papovaviruses contain, in addition to DNA, 3 internal polypeptides of which two may be cellular histones [208]. Arbovirus nucleocapsids, comprising one protein only and virion RNA [3], have also been termed 'cores' [648]. The complex 'cores' of oncornaviruses may or may not consist of a symmetrical capsid plus RNA and internal proteins (sect. VI).

In the case of enveloped viruses with helical symmetry, the nucleocapsid is also referred to as the core. This contains RNA, the capsid polypeptide and several associated proteins which may have enzymatic functions, i. e. polymerase activity as in the cases of the myxo-, paramyxo- and rhabdoviruses [650]. The *in vitro* and *in vivo* formation of cores from intact virions will be discussed in section VI.

### 2. Structures Recognized by Receptors

The specific virion structures involved in attachment of a few model viruses will be discussed as the data permit.

Figure 2 shows the outline of the capsomeres of adenovirus. The vertex or penton capsomere comprises an elongated antenna-like structure, the fiber, and a penton base; it is surrounded by 5 peripentonal hexons [222, 689]. These hexons differ from the hexons on the facets of the icosahedron in only one respect: one of their 6 neighbors is the penton [543]. The elongated fibers, which vary in length between different adenovirus types [493], constitute the attachment structure [533]. They are probably com-

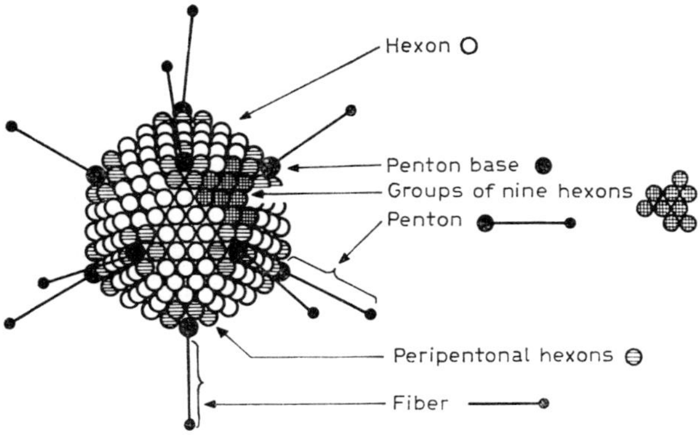


*Fig. 2.* Schematic drawing of an adenovirion showing the major components of the outer capsid.

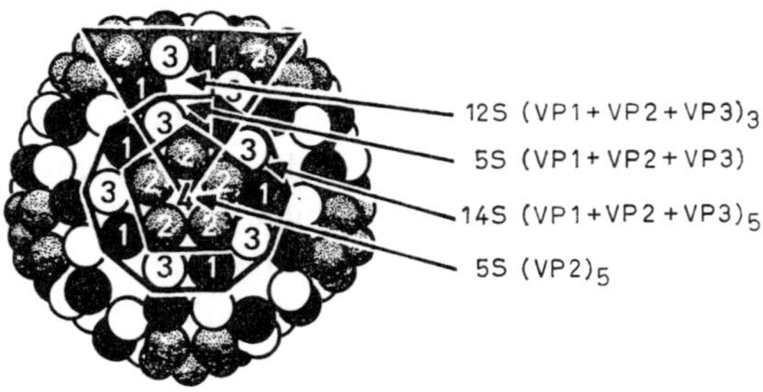


*Fig. 3.* A hypothetical structure of picornaviruses. View of the virion particle along the 5-fold axis showing the individual polypeptides VP1, VP2, VP3 and VP4 marked by numbers. The cleavage planes for the 14S and 5S degradation structures observed in EMC virus [175] and the 12S structure observed in FMDV [75] as well as the 5S structure comprising 5 VP2 observed in coxsackie B3 [527] are indicated. The 5 polypeptides of VP4 at the vertices are only labeled with a single 4 in the figure. This figure is modified from figure 12 in PHILIPSON *et al.* [527].

posed of 1–3 identical polypeptides which differ from those of the penton base and the hexon [655, 699]. It has also been suggested that the fiber polypeptides are phosphorylated [580] and contain sugar residues [310].

The attachment function has been ascribed to the fiber mainly because the structural proteins of this virus are produced in excess during lytic infection [726] and the free fibers have been highly purified and shown to inhibit attachment of the virion [533]. It can therefore safely be concluded for this virus that the vertex area of the virion and in particular the fiber is recognized by the plasma membrane.

The protein complement of picornaviruses consists of 4 polypeptides named VP1 to VP4 in order of decreasing molecular weight [18]. Two of these polypeptides (VP2 and VP4) are probably cleaved during maturation from the precursor VP0 [312, 313, 411]. Both $T = 1$ and $T = 3$ structures have been proposed for the picornaviruses [527, 578] and the former involves 12 pentamers, each containing all four polypeptides. The $T = 3$ structure postulates 20 hexamers composed only of VP1 and VP3 and 12 pentamers containing VP2 and VP4; this is illustrated in figure 3. It is also possible to construct a $T = 1$ structure with VP2 and VP4 in the vertices.

It has been suggested that the VP4 structure of picornaviruses is recognized in the early attachment at the plasma membrane [65]. This was based on the finding that loss of VP4 is correlated with alteration of antigenic character [64, 333] and also the fact that the first step of irreversible alteration involves loss of VP4 from the virion [127, 399]. However, irreversibly altered virus which has eluted abortively from the cell shows a lighter density than intact virions and it has been suggested that receptor structures are bound to it [126, 200]. Furthermore, it has been suggested that sulfhydryl groups are involved in the attachment step of picornaviruses [531], but VP4 lacks cysteine residues [527]. For these and other reasons discussed in section VI, it seems probable that if a single protein moiety is responsible for recognition of cellular receptors, VP4 cannot be the one. It is, however, not yet clear that a single polypeptide is recognized by the plasma membrane; the entire conformation of the virion's surface may also be the recognition unit. The fact that the receptors differ between picornaviruses suggests considerable specificity in the recognition mechanism (sect. III).

A membrane component of enveloped viruses appears to be the structure recognized by the plasma membrane. The proteins of the orthomyxo- and paramyxoviruses have been discussed above. The receptor-specific structures correspond to the hemagglutinin spikes which are composed of glycoprotein. Rhabdoviruses contain 5 virus-specific polypeptides [327,

467, 663, 705]; one of these, a glycoprotein, is needed for both attachment and hemagglutination and it is located in spikes on the virion envelope [80, 81]. The glycoprotein also constitutes the major neutralizing antigen [327]. Unlike the myxoviruses, the rhabdoviruses do not contain neuraminidase and the virion glycoprotein and host-specified glycolipids both contain sialic acid [348]. The glycoprotein-containing envelope of the arboviruses is also required for attachment to cells [55]. It is tempting to postulate that polypeptides involved in the attachment of most enveloped viruses will be shown to be glycoproteins.

## 3. The Ratio of Infectious to Physical Particles

Most animal viruses, unlike bacteriophages, require between 10 and 1,000 particles to score one infectious unit by plaque or other biological assays for infectivity [235, 605, 716]. There is, however, a linear relationship between the number of infectious units detected and the dilution of the inoculum, and this establishes that only one particle is required for the entire infectious cycle [406]. The high ratio of physical to infectious particles cannot, therefore, be attributed to a requirement for multiple infection. The high ratio might be due to the presence of a large proportion of inherently defective particles, or to inefficiency in any one of the following steps: attachment, irreversible alteration, penetration, uncoating, or the initiation of biosynthetic activities. Examples of all these mechanisms are available. The paramyxoviruses require about 40 particles for 1 PFU [Choppin, private commun.] and the virions are labile and may deteriorate during purification or handling. A low efficiency in penetration has been established for poliovirus where 60–70% of attached virus elutes as non-infective particles at 37 °C [321]. With both poliovirus and rhinovirus it is postulated (sect. VI) that many virions are also destroyed in lysosomal vesicles and this may also help explain the high ratio of physical to infectious particles. In the case of the adenoviruses, a virion must penetrate the plasma membrane and traverse the cytoplasm, and then the parental genome must establish itself in the nucleus of the cell [401]. Not more than 30% of the adenovirus genomes reach the nucleus and thus a factor of 3 between particles and PFU can be accounted for by deficiencies in these intricate steps. However, the ratio of physical to infectious particles is 10–20 for adenovirus type 2 [235] and thus a factor of about 5 must be accounted for by other defects.

## *B. The Structure of the Plasma Membrane*

We will discuss some recent developments which have influenced our view of the structure of the plasma membrane as it participates in the early events of virus-cell interaction. The compositions of the proteins and glycoproteins [223, 237, 259, 301, 641, 690] and lipids and glycolipids [223, 382, 690] of the cellular membranes have recently been reviewed elsewhere in detail.

### 1. The Components

Plasma membranes contain diverse species of proteins and there is no specific fundamental 'structural protein' subunit. The proteins also vary over a wide range in molecular weights and in their relative abundances; to some are also attached short heteropolysaccharide chains. One of the major glycoproteins of human erythrocytes has a molecular weight of 55,000 and it carries at least four unique carbohydrate residues which are released as glycopeptides after digestion with trypsin. Cyanogenbromide-mapping suggests that a number of each of these are located towards the N-terminal end, which presumably projects outside the membrane [430].

Lipids comprise roughly half the weight of cellular membranes; these are mainly cholesterol, triglycerides and phospholipids. A smaller but important portion contains a variety of glycolipids, mainly sphingoglycolipids. These include cerebrosides and globosides which contain only neutral sugars, and gangliosides which all contain N-acetylneuraminic acid (NANA) as well. The carbohydrate chains of glycoproteins and glycolipids are built stepwise by the action of specific glycosyltransferases; both lipids and proteins may have common carbohydrate structures and thus common antigenic determinants. For example, blood group determinants may reside on both glycoproteins and glycolipids. Since the structures of the complex carbohydrates result from sequential addition of monosaccharides, they generally show microheterogeneity dependent upon the extent to which every possible addition has been made to every available glycosyltransferase site.

Comparison of the glycoproteins and glycolipids of dividing and non-dividing cells or of transformed and nontransformed cells shows changing patterns of glycosylation. No consistent pattern relating to transformation and cell growth has been confirmed for the composition or size of glycopeptides released from cells by proteolysis [70, 71, 259, 448, 582, 713]. A difference has also been detected in the glycopeptides from the membrane of transforming as opposed to nontransforming avian oncornaviruses; this difference was attributed to the host cell in which virus was prepared, i. e. all avian oncornaviruses had equivalent glycopeptides when grown in transformed cells [376]. In addition to glycoproteins which are integral parts of cell membranes, there are superficial glycoproteins or glycopeptides (acid mucopolysaccharides) found at the outer surface of the cell; these have also been cited to increase after transformation [588, 610].

Glycolipids can be separated by thin-layer chromatography and significant alterations of these moieties have been related to the growth state or state of transformation of the cell [61, 133, 242, 243, 561, 583]. In general, the sphingoglycolipids of dividing and transformed cells show a decrease in the number of carbohydrate residues relative to those of confluent or nontransformed cells. Unfortunately, there is great variation in the pattern from one cell line to another, and in one study, the clone showing the least contact inhibition showed the most complete set of glycolipids, i. e. there is no correlation between ultimate cell density and surface structure [583]. However, if carbohydrate moieties of the cell surface should influence virus attachment, it is easily anticipated that cell lines will differ greatly from each other in receptor activity.

The basic structure of the membrane of the animal cell involves a bilayer of lipids oriented with nonpolar groups inwards as reviewed by SINGER [623]. The membrane may be described as a random mosaic of proteins and glycoproteins imbedded in the continuous lipid bilayer [623, 690]. The proteins contribute substantially to the thickness of the membrane which is 7–10 nm as seen by electron microscopy; lipids alone form membranes about 5 nm thick, and the difference probably results from the projection of portions of the proteins above or below the bilayer.

The dispositions of the proteins and glycoproteins have been investigated by controlled proteolysis or chemical labeling [708]. Physical and chemical evidence suggests that the proteins retain a globular structure [314] and that they contain hydrophobic regions of low polarity [78] which remain imbedded partly within the lipid bilayer. The hydrophilic portions of the proteins project from the bilayer outward, inward or in both directions relative to the cell [65a] and their outward or inward orientation is fixed [640] and cannot be altered without a prohibitive expenditure of free energy [623]. The $\alpha$-D-glucosyl or $\alpha$-D-mannosyl carbohydrate residues which react with the plant lectin concanavalin A (Con A) [102] reside almost exclusively on the exterior portion of the erythrocyte membrane [485].

Similarly, up to 74% of the entire sialic acid of mouse $TA_3$ cells can be released from the external surface by tryptic digestion of intact cells [103], and in general, a sizable portion of the cell's carbohydrate can be cleaved from the external surface by proteolytic enzymes [416, 609]. Some methods for the preparation of membrane fragments can produce either inside-out or rightside-out vesicles, and these may then have very different surface properties [640, 708]; this might be reflected by differences in virus receptor activity of membrane fragments produced under different conditions.

## 2. The Fluid Mosaic Model

It has recently become apparent that the components of the cell membrane are usually free to diffuse within the lipid layer; this leads to what has been termed the 'Fluid Mosaic Model' [625, 626]. By this view, the membrane is largely a two-dimensional liquid at physiological temperature. Some regions may, however, be in a crystalline state [624]. Spin-label studies indicate that a phospholipid can diffuse 5 μm/sec in the sarcoplasmic reticulum membrane at 37 °C [589]. The diffusion of larger subunits such as glycoproteins must be slower and the earliest evidence for the fluid nature of the plasma membrane was obtained by the microscopic observation of the gradual mixing of host antigens after fusion of human and mouse cells [212]. This was observed with the aid of fluorescent antibodies to human or mouse antigens and required about 40 min for completion, i. e. the antigens diffused about 5 μm/40 min. The rate of mixing was also dependent upon the temperature and fell by a factor of about 10 if the temperature was reduced from 37 to 15 °C [212]; this is in keeping with the viscosities of biological lipids in this range.

Other evidence for the fluid mosaic model has come from the ability of bivalent antibodies to attach to and cross-link immunoglobulin molecules located on the surface of lymphocytes [159, 519, 672]. There appear to be about $2 \times 10^4$ such specific immunoglobulin-receptor sites per lymphocyte [741]. The cross-linkage of the membrane-bound immunoglobulins with fluorescent anti-IgG results in the coalescence of distinct patches or flecks on the cell surface which eventually coalesce to produce 'caps' at one end of the cell, and are then ingested by phagocytosis. Patch formation is temperature-dependent but insensitive to metabolic inhibitors and thus appears to depend upon diffusion of the receptor units [159]. Not all multivalent ligands cause patch formation; Con A which binds extensively to the outer surface of cells produces no patches on normal lymphocytes [741], erythrocytes [485] or fibroblasts [482]. There are between $1.4 \times 10^6$ and $3 \times 10^7$ binding sites for Con A per cell [38, 307, 511, 741] and the effect of excess Con A appears to be to increase the rigidity of the lymphocyte membrane and prevent patch formation [741]. The effect may be due to extensive cross-linkage which restricts the mobility of surface components. However, with both transformed and trypsin-treated cells, Con A produces at least small patches which are visible in the electron microscope [482, 484, 573]. The ability of Con A receptors to coalesce may be the basis for the widely noted agglutination of transformed cells or trypsin-treated cells, but not normal cells, by plant lectins. The absolute number of Con A binding sites is not greatly altered upon transformation or trypsin treatment [307, 482, 511, 573] and it has been suggested that the increase in agglutinability results from the increased local concentration of receptors in patches, which increases the strength of the cell-to-cell bonding [482, 483, 626].

Some of the subunits of the erythrocyte membranes can be visualized in the electron microscope and are found distributed randomly over the cell surface. Treatment of the membranes with trypsin causes loss of surface carbohydrates, as glycopeptides, and probably also causes a partial loss of the polar regions of some membrane proteins which project in an exposed fashion from the lipid layer. This results in the aggregation of the visible morphological units so that patches are now detected in the electron microscope [431, 630]. SINGER and NICOLSON [626] have concluded that the remaining membrane proteins may still retain a large fraction of their original oligosaccharide chains but that the loss of some of the more polar structures is sufficient to permit the now more hydrophobic structures to cluster into patches. Such clustering may also be dependent upon the attraction of ionic pairs since it has been noted that the morphological units of nontrypsin-treated human erythrocytes also cluster at low but not physiological ionic strength at pH 5.5 [151].

Thus it is obvious that the state of aggregation of the units of the cell membrane can be altered by multivalent ligands, by chemical alteration of the hydrophobicity of the subunits, and possibly by pH and ionic strength. The state of aggregation of the subunits may also affect cell-cell interaction. Finally, the rate at which subunits can aggregate within the cell membrane is dependent upon the viscosity of the cell membrane and hence upon temperature. In section III we will attempt to show that the ability of receptor subunits to coalesce at the site of virus attachment may be crucial for the first step of infection.

This discussion has not adequately emphasized the unusual mobility of the surface of the fluid membrane. Time-lapse cinematography has demonstrated that the plasma membrane of cultured cells is constantly involved in endocytosis and exocytosis [WESTERMARK and PONTÉN, private commun.] and in this way the distinction between the inside of intracellular vesicles and the extracellular surface is also blurred.

### 3. Virion Envelopes

One area of uncertainty with the fluid mosaic model concerns the nature of the forces which hold the membrane glycoproteins of enveloped viruses together during the process of budding from the infected cell. Electron microscopic investigation has shown that the neuraminidase units of the myxoviruses or paramyxoviruses are not free to diffuse during budding [349]. In this regard, the surface above a budding virion may be viewed as a patch of virion glycoproteins, but the nature of the forces responsible for holding them together are not yet known, nor is it known whether or not the virion glycoproteins are free to diffuse in the virion envelope. In general, the envelopes of viruses are formed by insertion or addition of virus-specific glycoproteins or proteins into or onto the cells' lipid bilayer and the lipid composition of the virions resembles that of the host cell [346, 347, 445, 550]. The proteins are, however, virus-specific but the carbohydrate portions of the proteins [73] and glycolipids [346, 348] are synthesized mainly with the enzymes of the host cell. The glycosylation of the specific virion proteins may provide a useful system for the study of the formation of complex polysaccharides. Some evidence has recently been found that the glycosylation of the hemagglutinin protein of influenza occurs sequentially and for the most part away from the microsomal site of synthesis [109, 637]. The process by which the fully glycosylated protein is transported to the outer surface of the cell is not known.

## *C. The Identity of Virus Receptors*

### 1. The Receptor Concept

A virus receptor is a natural component which binds virus under physiological conditions. This concept is imprecise because it does not define the magnitude of the binding constant and receptor monomers may bind virus only very weakly unless coupled into polymeric structures (see below). Many methods are used to detect binding. The simplest employs intact cells from which suspended virions can easily be removed by washing; the extent of binding is then determined by infectivity assay, hemagglutination assay, or by counting radioactively labeled virions. When broken cells are employed, the small fragments cannot easily be washed free of virus, but the complex may impart an altered sedimentation rate, buoyant density or electrophoretic mobility. Other methods include the detection of a decrease in the infectivity or hemagglutination of bound virus and these depend upon a competition between subcellular and cellular receptors for the virions. An interesting variation has been applied in studies with the bacteriophage R17, in which case the receptors are located upon 'pili' filaments which can be absorbed on cellulose acetate filters together with any attached radioactive virus [145]. Variations based upon this principle might be applied in animal virus systems.

The concept of a receptor is also imprecise because it represents a structural configuration for which the virion shows affinity; this configuration is usually a part of a large molecule, but portions of several molecules might together also provide the required specificity. Receptors for different serotypes of virus may also partially overlap and have both individual and common properties. Thus the receptor may not consist of a single chemical entity, and is loosely analogous to an 'antigenic determinant'. There may be more virion receptors per cell than 'receptor sites'; the latter are determined experimentally by the number of virions or of specific virion subunits bound per cell. This approach showed that there are 10–100 times more receptor sites per cell for adenovirus type 2 fiber antigens than for the corresponding virions [533]. Both recognize the same sites and the numerical nonequivalence can be attributed either to the binding of the small fiber proteins to sites inaccessible to virions for steric reasons, or to the possibility that the virion can bind receptors to each of its 12 fibers.

The attachment of adenovirus type 2 fiber illustrates that smaller entities than intact virions can attach to host cells. Many hormones probably interact with specific receptors at the cell surface and recent work shows that fat cells have about $10^4$ receptors for insulin [130]. Cuatrecasas [131, 132] has employed affinity chromatography for the purification of these receptors and this approach may eventually be applied to receptors for animal viruses.

## 2. The Cellular Location of Virus Receptors

A virus receptor must function at the cell surface but intracellular receptors may also exist. The receptors of paramyxoviruses, myxoviruses, and certain other viruses contain sialic acid and they are *a priori* likely to be found at the cell surface together with the majority of cell-associated sialic acid [103, 609].

Electron microscopic examination of the interaction of influenza virus with isolated plasma membranes has shown that virion attachment occurs on the extracellular side only [100]. The receptors for several other viruses have been shown to reside primarily at the outer surface of the cell membrane. When proteolytic enzymes were used to destroy the receptor activity of intact cells for poliovirus type 1 and coxsackievirus B3, no intracellular receptors were detected after the treated cells were washed, broken and tested for ability to inactivate virus infectivity [444, 749]. It is unlikely that the proteolytic enzymes penetrated the intact cells. Earlier findings that microsomal fractions of cells contained enterovirus receptors may be attributed to the presence of fragments of the plasma membrane in the microsomal fraction [277, 528].

The receptors for adenovirus type 2 can be detected in fragments of broken cells, by taking advantage of the low buoyant density of the virion-receptor complex in CsCl gradients. Receptors on intact HeLa or KB cells could be destroyed by treatment with subtilisin, whereafter few intracellular receptors could be detected in the broken cells. It was also ascertained that subtilisin had not entered the cells because a subtilisin-sensitive cytoplasmic enzyme was not inactivated during treatment [533].

As already discussed, much of the cellular carbohydrate is found at the exterior surface of the plasma membrane. This reflects the general function of glycosylation of proteins which is to regulate their extracellular distribution within the organism [736]. The surface carbohydrates provide specific sets of structural or antigenic surfaces to various tissues and this probably contributes to the specificity of virus receptors.

## 3. The Regeneration of Virus Receptors

Since most cellular receptors are probably at the cell surface, it has been possible to examine the rate at which they are regenerated after enzymatic removal. The sialic-acid-containing receptors for paramyxoviruses appear to be half regenerated within 6 h incubation after inactivation by neuraminidase [433]. Other studies reveal that the rate of incorporation of membrane-bound sialic acid is linear and constant for both enzymatically treated and untreated cells [302]. Since removal of receptors does not stimulate their regeneration and because many molecules in addition to those serving as receptors for a particular virus may also be removed, it is *a priori* unlikely that enzymatic treatment of cells will enhance the radioactive labeling of a specific set of virus receptors.

The rate and kinetics of regeneration of receptors varies from one virus to another. Adenovirus type 2 receptors required about 6 h for half regeneration after cells were treated with subtilisin [533]. Poliovirus receptors required 1–2 h for regeneration after removal with trypsin while coxsackie B receptors required significantly longer after removal with chymotrypsin, with full regeneration only after two days [394]. These differences in rates of regeneration of receptors indicate that the synthesis of specific and discrete biochemical units are involved in virus attachment. Sometimes cells from fresh tissue have been found to be unable to adsorb virus and this may be a result of the use of trypsin for their dispersion. For example, primary calf kidney cells were only able to adsorb and eclipse a bovine enterovirus after 24 h of culture [88].

## 4. Receptors *in vitro*

It has been possible to demonstrate the attachment of virus to host cell receptors in a few subcellular systems. In addition, some competitive inhibitors of virus attachment are derived from cellular receptors (sect. III). Erythrocyte receptors are often easier to prepare than host cell receptors because erythrocyte stroma is easily obtained in quantity. (Erythrocyte receptors for myxoviruses, adenoviruses and for certain enteroviruses are discussed primarily in section IV.) There may, however, be differences between receptors from host cells and erythrocytes, and hemagglutination inhibition may therefore not always be a suitable assay for use during fractionation of host receptors. Many viruses also do not hemagglutinate.

The rapid interaction of influenza virions with small subcellular vesicles detached from normal cells was studied by Hoyle [293], and influenza virions also attach to chick embryo cell plasma membranes [364] at 0 °C as shown by electron microscopy [100]. A variety of enveloped and nonenveloped viruses are also inactivated by incubation with disrupted cells

[547]. A receptor for influenza virus was purified from erythrocyte membranes as will be discussed in section IV; the chemical properties of this material have been detailed and it is also known to carry M and N blood group specificities. These or closely related antigens are probably also distributed on other cells which are susceptible to influenza virus infection [634]. The receptors are mainly glycoproteins which appear to be free to diffuse within the plane of the cell membrane [431] and which have folded $\alpha$-helical regions, i. e. they are not spread out as a sheet on or in the cell membrane [314, 634]. The receptor specificity can be attributed to the carbohydrate side chains; myxoviruses also attach to other macromolecules containing terminal sialic acid residues. This property was used to reconstruct artificial membranes containing receptor units in the form of the glycoprotein fetuin which attached influenza virus and Newcastle disease virus (NDV) but not adenoviruses or herpesvirus [680]. Some gangliosides also contain the correct carbohydrate structure to serve as myxovirus receptors (sect. IV).

Adenovirus type 2 has been shown to attach by specific receptors to cell membrane fragments and to partly purified plasma membrane vesicles; virus-receptor interaction was detected by a shift of the buoyant density of virions after attachment of the membrane fragment [533]. In an erythrocyte system, employing adenovirus type 7, it was possible to detect receptor-virus interaction by a small decrease in virus hemagglutination [476] a method which in practice proved inaccurate [NEURATH, private commun.]. In either method, there is probably a practical limit to either the size of the membrane fragment or the state of aggregation of the receptor unit during assay and more sensitive methods of detecting virus-receptor interaction will be needed for further progress.

Arboviruses of group B interact with subcellular fractions from susceptible cells as well as fowl erythrocytes, as revealed by hemagglutination inhibition and infectivity inhibition [461, 470]. An 'active fragment' was extracted by phenol treatment of partly purified material from bovine brain: this unit had a molecular weight of only about 1,000 [746]. The 'fragment' contained mannose, glucose and traces of galactose and ribose plus a small amount of a ninhydrin-reactive material. A partly purified brain receptor containing the active fragment plus lipids was able to attach Japanese encephalitis virus at physiological temperatures and the reaction was not reversible by 5 M NaCl, while the reaction with the active fragment alone was reversible. However, active fragments mixed with artificial membranes produced from lecithin and cholesterol *in vitro* (liposomes) gave irreversible attachment similar to the original receptor-lipid complex [744]. Mixtures of various complex lipids were also able to interact with group A or B arboviruses as detected by hemagglutination inhibition [211, 230]. Phosphoinositides have been singled out as particularly active substances [211, 211a], but the relationship between these materials and the defined mixture of 'active fragment' and lipid from bovine brain remains unclear. In conclusion, a mixture of lipids, plus a glycolipid or a glycopeptide, can interact strongly with certain arboviruses but the relationship of this interaction to the attachment of virions to host cells is uncertain.

Many picornaviruses interact with receptors *in vitro*. The enteroviruses have been studied in greatest detail and the erythrocyte receptors for the hemagglutinating enteroviruses have been solubilized. As already mentioned above, picornavirus receptors on host cells are regenerated after digestion of intact cells with proteolytic enzymes and they are probably elaborated continuously in growing cells. Such receptors might therefore be displaced from the cell and free receptors for various enteroviruses can be found in the supernatant of cells grown in serum-free medium [597]. This material has been detected by hemagglutination inhibition tests and may be a lipoprotein. However, it is also produced from nonprimate cells which are unable to attach polioviruses and this constitutes an important difference from the receptors found in lysates of host cells or erythrocyte membranes [597].

The receptors for poliovirus can be detected in broken host cells by their ability to inactivate virus infectivity [27, 276, 277, 442, 547]. The interaction of poliovirus with membrane fragments has also been detected by alteration of the electrophoretic mobility of the virus [677]. Receptors for coxsackieviruses have also been detected *in vitro* [443, 564, 749]. Attempts to solubilize and purify enterovirus receptors from host cells have not progressed rapidly but some success has been reported by McLaren *et al.* [444] who used 1 % triton X-100 detergent plus ultrasonic treatment. The soluble material contained receptors to viruses of different receptor families, and poliovirus receptors were present only in extracts from cells which were susceptible to poliovirus infection. The poliovirus receptor was partly purified by CsCl gradient centrifugation and had properties of a complex of glycoproteins and lipoproteins. Rhinovirus receptor activity could not be solubilized from host cells by this technique, but the bond formed *in vitro* between type 2 rhinovirus and the receptors on membrane fragments from susceptible cells is weaker than the poliovirus-receptor bond; this implies that new and sensitive methods will be required to detect rhinovirus-receptor interaction *in vitro* [Lonberg-Holm, unpublished].

# III. Attachment of Virions to Host Cells

## *A. The Specificity of Receptors*

The specificity of virus-receptor interaction will be documented in this section.

### 1. The Number of Receptor Sites per Cell

Viruses attach to a limited number of specific receptors per cell and the number of attaching virions is not determined by the total surface area of the cell and the cross-sectional area of the virions. It was found, for example, that the amount of coxsackievirus B3 attached to HeLa cells did not increase linearly with increasing input virus [123] and the receptor sites appeared to be saturated after about 200–300 PFU attached per cell, corresponding to $1 \times 10^5$ particles per cell since coxsackie B3 shows a ratio of 100–500 particles/PFU [126]. Other more recent work in the laboratory of one of the authors has examined the attachment of several radiolabeled picornaviruses and has also shown saturation of host cells by different viruses and at different saturation levels [399]. Figure 4 shows the adsorption of poliovirus type 1 by HeLa cells after 15 min of incubation. Other work has shown that polio types 1, 2 and 3 share the same receptors [123; CROWELL, personal comm.].

The adsorption isotherms for poliovirus (fig. 4) and for other picornaviruses do not show complete saturation at high input multiplicities. This incomplete saturation gives only an approximate number of receptor sites per cell but poliovirus virions attach with low efficiency after more than $3 \times 10^3$ have attached per HeLa cell. Similarly, partial saturation of rhinovirus receptors can be demonstrated at $1\text{–}2 \times 10^4$ particles per cell. The incomplete saturation may be due to heterogeneity of the affinities of receptor sites for virions. Irreversible alteration or eclipse of virions during the attachment reaction does not influence the shape of the adsorption isotherm since the addition of an inhibitor of the eclipse reaction did not alter the

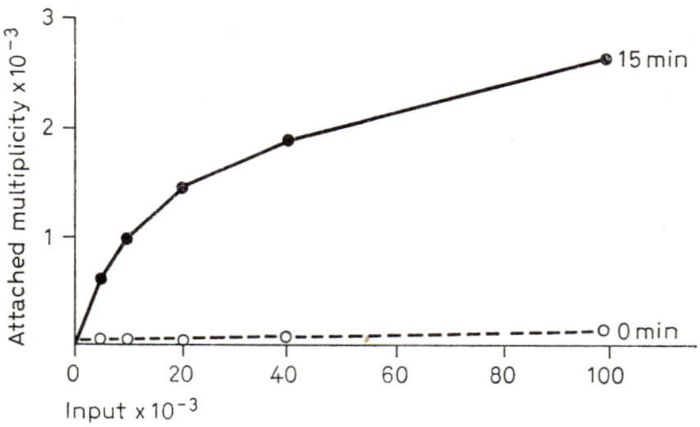


Fig. 4. Effect of input multiplicity on attachment of poliovirus type 1 (Mahoney) to HeLa cells grown in suspension culture. Suspensions of $10^7$ cells per ml, and purified tritium-labeled virions up to $10^5$ particles per cell, were mixed at 0 °C and incubated for 0 (O) or 15 min (●) at 37 °C. The cells were washed and the number of virions attached per cell was calculated from the cell-associated radioactivity [LONBERG-HOLM, unpublished].

curve for rhinovirus type 2 [NOBLE-HARVEY and LONBERG-HOLM, in preparation]. The area covered by $10^4$ picornavirus particles is only about 1 % of the plasma membrane of a HeLa cell [399] and hence the saturation does not result from a spatial limitation. Apparently differing results have been obtained for the number of picornavirus receptor sites of fibroblasts grown and infected as monolayers: these were found to have about 10–100 times more receptor sites than described above for HeLa cells [446]. The two systems will require further comparison.

There are about $10^4$ receptor sites per cell for adenovirus type 2, but a 10- to 100-fold greater number of adenovirus type 2 fiber antigen molecules can attach [533]. These fibers appear to recognize the same receptor as virions and the difference in saturating levels has been attributed to steric factors or to polyvalent binding of virions. The free fiber protein of certain adenoviruses is produced in great excess in infected cells and can react stoichiometrically with the receptors for a number of adenovirus serotypes. It now appears likely [533] that it is the excess of fiber which is the active component of the 'erythrocyte receptor-modifying factor' [331, 332] of infected cells.

No evidence for the saturation of cells by myxo- or paramyxovirus is available. It seems probable that virions which attach rapidly to many types of sialic acid-containing molecules will have many more receptor sites than virions which attach more specifically; the former may even be limited by the space available on the cell surface. The attachment to erythrocytes of influenza virions which had been heated to destroy neuraminidase activity, and of polyoma virus (which contains no neuraminidase but which also recognizes sialic acid-containing receptors) have been compared [462]. This study indicated that there are fewer polyoma receptor sites than influenza sites and that influenza virus recognizes polyoma receptors and also receptors not recognized by polyoma [462]. The data do not permit a calculation of the number of polyoma sites per cell.

## 2. The Concept of Receptor Families

Since the receptors for many viruses can be saturated by excess virus, it is possible to test whether two viruses compete for a common receptor. Two viruses which share the same receptors belong to the same receptor family. Membership can, however, be partial, i. e. a virus can belong to two receptor families or there may be a minor amount of cross-interaction. Thus, influenza virus could saturate the erythrocyte receptors for polyoma virus but polyoma could only saturate some of the receptors for influenza.

Evidence for receptor families among nonenveloped viruses was first found by exposing HeLa cell fragments to formolized type 1 poliovirus (vaccine) [548]. The attachment of infective poliovirus of all 3 serotypes was reduced while the attachment of coxsackie B viruses or of NDV and equine encephalitis viruses was not affected. Attachment was measured by a decrease in infectivity after incubation with the cell fragments. Later work with intact cells and noninactivated virus showed that excess coxsackievirus B3 and coxsackievirus B1 competed with each other for receptors as did poliovirus types 1 and 2; however, coxsackievirus B3 did not interfere with the attachment of poliovirus type 2 and vice versa [123]. Attachment of virus was measured by the decrease in PFU from the supernatant after neutralization of the virus used in excess for blocking. The conclusion that poliovirus and coxsackie B viruses are members of different receptor families has been confirmed with purified radioactive virus [Crowell and Lonberg-Holm, unpublished]. Cells chronically infected with certain coxsackie B viruses were unable to attach other coxsackie B viruses but could be superinfected with coxsackie A viruses or poliovirus [122, 128]. This finding might be explained by the saturation of the receptors of the coxsackievirus B receptor family. The saturation of erythrocyte receptors for echovirus 11 and 19 and coxsackievirus B3 by prior treatment with echovirus 7 indicates that all four of these viruses share the same receptor, but these data require confirmation [528].

Rhinovirus type 2 can attach to monkey cells while type 14 cannot [240] and hence the receptors are separate genetically determined units. Experiments with purified radioactively labeled rhinoviruses have also shown that there are at least two separate rhinovirus receptor families [399; LONBERG-HOLM, unpublished results]. The first family contains types 1A, 1B and 2 while the second contains types 5, 14, 41 and 51; the first probably corresponds to the M strains which can be propagated in simian cells and the second to H strains which require human cells. The receptors of these two families differ from each other and from the receptors for polioviruses and coxsackie B viruses [399; CROWELL and LONBERG-HOLM, unpublished].

Any adenovirus whose attachment is inhibited by adenovirus type 2 fiber antigen must belong to the same receptor family as adenovirus type 2. So far, this family definitely includes only adenovirus type 5 [533] but it probably also includes the adenoviruses which are sensitive to the 'erythrocyte receptor modifying factor' [331, 332]. Adenovirus fiber antigen does not inhibit poliovirus uptake and hence polioviruses belong to a distinct family.

### 3. Organ and Species Specificity of Receptors

This subject was thoroughly reviewed 9 years ago [273, 367] and there is little to be added. It has, however, become increasingly clear that host range can be determined by other factors than the presence of receptors (sect. V) and receptor specificity has therefore been placed in perspective.

Species-dependent receptor specificity occurs, however, among human picornaviruses, many of which attach to and multiply in only the cells of primates. Erythrocytes of primate origin are also required for agglutination by echoviruses 7, 11 and 19 and for coxsackievirus B3 [528]. Poliovirus receptors are found in susceptible tissues of primates but not in insusceptible tissues, such as heart or skeletal muscle [27, 270]. Primary or established cultures of primate cells which were susceptible to polio infection did show specific attachment of virus while all of a variety of cultured non-primate cells were insusceptible and also did not attach virus [442]. One exception appeared to be an established 'embryo rabbit kidney line' [442, 677] but this was later shown to consist of contaminating HeLa cells [62]. It was suggested that the basis for attenuation of poliovirus could be due to an alteration of the virion's receptor specificity, but careful examination of attachment of attenuated and neurovirulent poliovirus to primate nervous tissues revealed that the rates of attachment were the same [254, 366].

In addition to primate cells, coxsackieviruses must be able to attach to and infect mouse cells since they are characteristically pathogenic for newborn but not adult mice. KUNIN [366, 367, 369] detected receptors for coxsackie B1 and B3 in homogenates of the tissues of newborn mice but these decrease in concentration during maturation. HeLa cells contain receptors for coxsackie B viruses, for which they are permissive [126, 443] but nonpermissive established mouse L cells do not contain receptors [126]. CAME and CROWELL [77] were unable to detect the presence of receptors for a representative group A coxsackievirus in minced fetal tissue or on cultured mouse cells. Coxsackievirus A9 infects primary human and monkey cells in tissue culture but does not infect HeLa cells or other established human cell lines which lack specific receptors [443]. Limited replication of coxsackie A viruses occurs in primary fetal mouse cells [225], in spite of the absence of detectable receptors. Differentiating muscle cells of fetal mice could, on the other hand, be infected with coxsackie A viruses [224, 377]; this correlates well with the fact that muscles of newborn mice are specific *in vivo* targets for coxsackie A virus. The susceptibility of the muscle cells was attributed to the appearance of specific coxsackievirus receptors on the myoblast cells before and during their fusion which leads to myotube formation. Mouse myoblast cells were not sensitive to infection by poliovirus during this process and this appears to rule out a nonspecific route of infection from the fusion process itself [224].

It is possible to select resistant variant cell lines from susceptible parent cells. Although some poliovirus-resistant cells may owe their resistance to specific inability to attach poliovirus [629], attachment is usually relatively normal but eclipse or uncoating is blocked [150, 271, 695]. When resistance is due to lack of receptors or an inability to uncoat the virion nucleic acid, the barrier to infection can be overcome by two methods. In the first, cells are infected with virus nucleic acid; nonprimate cells have been infected with poliovirus RNA [278] and primate cells lacking specific receptor have been infected with RNA from coxsackievirus A9 [443]. Secondly, transcapsidants or pseudotype virions may be produced in which the nucleic acid of one type of virus is enclosed in the coat or envelope of another virus for which the cell has receptors [274]. A transcapsidant of poliovirus RNA in a coxsackievirus B3 capsid was used to infect a resistant cell line which lacked poliovirus receptors [629]. Pseudotype virions containing heterotypic envelope proteins have also been produced with VSV nucleocapsids and envelopes of certain leukoviruses and these show an altered host range [751, 752].

## 4. Genetic Determinants

The antigen structures on the surface of the cell are of obvious importance for histocompatability and blood typing. Similar structures may also serve as virus receptors and hopefully the genetics and immunochemistry of each virus receptor will be related to that of an identified cell component. At the present time there is only sparse information upon which to establish such relationships. The M and N blood group specificities reside on glycoproteins which are identical with the major human receptor for influenza [335, 634]. In a similar way, a blood group antigen (isoantigen R1) may possibly serve as a receptor for or in some way influence the penetration of avian subgroup B leukosis-sarcoma viruses although the evidence is incomplete [121, 534, 642]. A relationship exists between alleles of the H-2 histocompatability locus of mice and susceptibility to tumors by murine leukoviruses but this is probably not due to an effect on virion-cell interaction [397, 675]. It has also been noted that type A human erythrocytes are more easily agglutinated than types 0 or B by reoviruses; this difference may possibly indicate that the isoantigens of the ABO alleles participate in reovirus attachment, although other explanations are also possible [69].

Heterokaryons between viral permissive and nonpermissive cells are often permissive [236] suggesting that permissiveness can be a dominant character. This has also been studied in hybrid cells where it may be possible to identify the chromosome controlling the receptor [236]. Nonprimate cells are not susceptible to poliovirus infection and susceptibility of human-mouse hybrids depends upon the presence of human genes. Permissiveness can be lost, probably by the loss of a human chromosome, and this loss correlates with the loss of poliovirus receptors [446, 712]. Such resistant lines can be found after infecting susceptible cells with one serotype of poliovirus; in each case the resistant cells are resistant to all three poliovirus types [370]. This cytogenetic approach may eventually aid in the identification of virus receptors and may also help construct biochemical and genetic maps of the various receptors, since cells deficient in a selected set of receptors can easily be tested with many other virus types [446]. It has recently been reported that treatment of hamster cells, which lack poliovirus receptors, with DNA of monkey cells produced transformants which were able to attach poliovirus [60]; unfortunately, this trait was lost after subcultivation.

## 5. Receptors during the Cell Cycle and after Transformation

The amount of the various enzymes and antigenic determinants of the cells' membrane fluctuate in a cyclic manner during the division cycle [56, 206, 676], and it may be expected that this variation will affect the early events in virus infection. Poliovirus replication has been studied in considerable detail in synchronized cells [187, 188] or in cells arrested in metaphase [39, 315]. The attachment of poliovirus to the host cells has, however, not been examined carefully, but it appears to occur normally, even during metaphase, and the variations observed in the rate of synthesis of virus-specific RNA are thought to reflect intracellular changes. Polyoma replication occurs preferentially in dividing mouse cells and there is a cyclic variation in susceptibility to infection but there is no evidence that this reflects variation in receptors [678].

Human lymphocytes are refractory to infection by many viruses, but after they are stimulated to divide by phytohemagglutinin (PHA) they serve as hosts for mumps virus [167], herpes simplex virus [469], yellow fever virus [23], VSV [180] and vaccinia virus [450]. With vaccinia virus, PHA appears to enhance the yield by acting at a stage subsequent to attachment. Mouse lymphocytes also replicate VSV and mengovirus more effectively after treatment with PHA [191]. PHA also enhances the replication of poliovirus in human lymphocytes and treatment increases the rate of attachment and also the rate or efficiency of the reaction leading to eclipse [729, 731].

Transformation of cells by oncogenic viruses alters the cellular membrane and may result in changes in the ability of the cell to interact with viruses. A line of polyoma-transformed BHK-21 cells could neither attach nor be infected by 4 strains of foot-and-mouth disease virus [160]. It is not clear whether the loss of foot-and-mouth disease receptors was caused by transformation or whether a cell line deficient in receptors was selected. After transformation by SV40 virus, hamster and human cells show increased resistance to herpes simplex infection [691] and human cells show increased resistance to poliovirus infection [241]; but the attachment of virus was not affected by the transformation. Hormones may also be able to affect the ability of cultured cells to attach certain viruses as suggested by results with herpes simplex [565].

### 6. Domains upon the Cell Surface

In addition to variation from cell to cell, there may be variation in the density of receptors from region to region on the surface of a single cell. This possibility is self-evident in cases of cells with unusual morphological differentiation but in other cases there may also be local 'hot' regions on the cell surface. Specific regions for preferential attachment were detected when mouse L cells were infected with VSV [261, 622] and chick embryo cells with leukoviruses [140] but this could be an artifact of fixation, or virions could migrate to such regions after attachment elsewhere on the cell.

### 7. Attachment with Low Specificity

Some viruses can attach nonspecifically to cell or other surfaces. Myxoviruses and paramyxoviruses interact with a wide variety of sialic acid-containing macromolecules and hence can be expected to have relatively low specificity for host or erythrocyte receptors. Cardioviruses, equine rhinoviruses and polyomaviruses also recognize a large number of sialic acid-containing receptors (see below) and also may be expected to show relatively low specificity in attachment.

Viruses can also attach nonspecifically to nonbiological surfaces. The attachments of fowl plague and vaccinia viruses to cells and to glass and aluminum surfaces have been compared [10, 688]. The aluminum and glass surfaces were able to adsorb virions at the maximum rate theoretically possible as calculated from the diffusion coefficient of the virions; this was more rapid than the uptake by cells. Influenza and adenovirus type 6 also attach to 'cabosil' powder and attachment of influenza was also shown to be inhibited by specific antibodies [46]. Two groups of coxsackieviruses have been distinguished by differences in their adsorption to bentonite [613], and also lipids such as hexadecylamine bind several enteroviruses [747]. The latter system showed interesting parallels with the reaction between enteroviruses and soluble receptors.

## *B. Inhibition and Enhancement of Attachment*

The specificity of virus-receptor interactions may be studied by selective inhibition which may provide clues to the nature of both the receptors and the virus structures responsible for receptor recognition.

### 1. Irreversible Inhibition of Virions

The conformational structure of nonenveloped viruses may be altered so as to destroy the ability of the capsid to react with receptors. Chemical alteration of virions may, however, have two effects, one specifically on a reactive chemical group and the other on the conformation of the virion. It is not always possible to decide between these alternatives when an altered property is observed.

Reagents which react primarily with amino acids have been employed to inactivate viruses. NEURATH *et al.* [477] tested adenovirus type 7 virions or dodecons with several reagents, and were able to inactivate hemagglutinating activity with 2,3-butanedione which reacts with arginine. The results were interpreted to suggest that arginine on the virion forms a salt bond to a negative charge on the erythrocyte membrane. 2,3-Butanedione had no effect on influenza virus hemagglutination and had no effect on the buoyant density of intact adenovirus virions. The hemagglutination of a paramyxovirus and a pseudomyxovirus (measles) have been inactivated with formaldehyde or $HNO_2$ under conditions which suggest that an essential free amino group has been blocked [475, 478].

A number of hemagglutinating enteroviruses are inactivated by reagents such as PHMB and oxidized BAL, which react with sulfhydryl groups of proteins [7, 72, 530, 531]. The hemagglutination of EMC virus has been reported to be inactivated at relatively high concentration of PHMB [368] but not at a lower concentration [7, 530]. The hemagglutinins of a simian enterovirus [260], of the 3 types of reovirus [72, 228, 388], of rubella virus [214], of a coronavirus [405] and of rabies virus [7, 371] have also been inhibited by PHMB. The infectivities of influenza viruses, NDV and arboviruses of group B are sensitive to PHMB [7], whereas their hemagglutinins are relatively inert [291, 405, 530]. PHMB and similar reagents also inactivate the infectivity but not the attachment of vaccinia and fowl plague virus [6]. The action of sulfhydryl reagents is usually reversed by excess sulfhydryl-containing compounds and hence they do not disrupt or alter the virions permanently [228, 405, 530]. These studies suggest that free sulfhydryl groups may be involved in attachment but that they probably do not participate to form a disulfide bond between virion and receptor. Attachment of poliovirus to HeLa cells is not inhibited by excess mercaptoethanol or other sulfhydryl-containing reagents but infectivity is destroyed by PHMB [271]. Thus the sulfhydryl reagents may also block early steps in infection subsequent to attachment [531].

Reagents have also been applied to virions in order to demonstrate the presence of carbohydrates. With the exception of glycosidases, however, these reagents also react with protein components. Periodate reacts with serine and other amino acids [101] as well as cis-glycols, while sodium borohydride reacts with disulfide bonds as well as carbonyl groups [390]. It was suggested [681] that carbohydrate might be present in the virions of hemagglutinating nonenveloped viruses by analogy with the myxoviruses or paramyxoviruses whose hemagglutinating subunits are glycoproteins. The ability of $KIO_4$ [388, 681], or $NaBH_4$ [390] to inhibit the hemagglutination of enteroviruses and reoviruses or of $KIO_4$ to inhibit EMC hemagglutination [368] cannot be taken as compelling evidence for the presence of essential carbohydrate within or on the capsid. Other evidence for the presence of carbohydrate has been obtained by treating virions with various glycosidases. Lysozyme reduced the infectivity of poliovirus type 1 and of echoviruses 7, 11 and 19 as well as the hemagglutination of reovirus type 2; a N-acetylglucosaminidase and other glycosidases also inactivated the hemagglutination of 5 echovirus types and of reovirus 2 [390]. The inhibition of enterovirus hemagglutination did not always correspond to a decrease in infectivity, suggesting that hemagglutination is not always a precise model for attachment. In studies of this type, it must be ruled out that the enzyme preparations do not contain proteolytic activity or a component which can combine with and inactivate the virions. The inactivation by glycosidases, and the ability of low molecular weight carbohydrates to interfere with the attachment reaction (see below), have nevertheless been evoked as evidence that the picornaviruses and reoviruses contain carbohydrates [216, 368, 387, 389, 390, 681]. So far, careful examination of both EMC virus [74] and coxsackievirus B3 [PHILIPSON and CROWELL, private commun.] have failed to detect virion-bound polysaccharides. Radioactive glucosamine has been detected in preparations of echovirus 12 [247] but in this case the evidence suggested that the carbohydrate may have been located in a cellular glycolipid bound to or entrapped within the virion; such material could conceivably function as a cofactor in the attachment of virus to cell and this possibility must be kept in mind when considering the evidence below for the inhibition of attachment by low molecular weight carbohydrates and glycolipids. Thus the existence of carbohydrates in nonenveloped hemagglutinating viruses remains in dispute.

For the enveloped viruses such as rhabdo-, myxo- and paramyxoviruses, it is widely documented that the envelope glycoproteins are involved in hemagglutination and attachment to cells. Proteolytic digestion in some cases cleaves the glycoproteins so that the carbohydrate portion is released from a smooth virion surface [80, 95, 113, 350]. The resulting glycoprotein-deficient particles lack not only hemagglutinin and neuraminidase, but have at least in some cases lost hydrophilic properties which may be needed to keep the virions in solution [95].

Purified, protease-free, N-acetylglucosaminidase from *Aspergillis niger* inactivates the hemagglutinin of PR8 influenza but not of NDV; inactivation of influenza coincides with the release of reducing sugar and hence is the result of a hydrolytic cleavage [40]. It seems likely that the carbohydrate portion of the surface glycoproteins functions in attachment. On the other hand, treatment of enveloped virions with periodate may not always readily inhibit hemagglutination or infectivity [371].

## 2. Irreversible Inhibition of Receptors

The receptors for different viruses differ in their sensitivities to irreversible inhibition by specific and nonspecific reagents. Poliovirus receptors in disrupted HeLa cells are inactivated at 56 °C [276]; at 60 °C both polio and western equine encephalitis virus receptors were inactivated while receptors for NDV and two types of coxsackie B viruses were not [548]. Erythrocyte receptors for echovirus 7 were inactivated at 60 °C but not at 56 °C; these receptors were shared by coxsackie B3 [528]. A careful study of receptors on intact HeLa cells showed a greater heat or acid lability of poliovirus receptors than of coxsackie B virus receptors; echovirus 6 receptors were similar to coxsackie B virus receptors in lability but the ability of heated cells to eclipse attached echoviruses differed from their ability to eclipse coxsackieviruses [750].

Receptors for many viruses in disrupted cells show loss of activity upon prolonged sonic treatment [260, 276, 528, 548]. Poliovirus receptors are sensitive to such inactivation, but controlled treatment can be employed to disrupt intact cells without a significant decrease in receptor activity [LONBERG-HOLM, unpublished].

The use of biochemical reagents which make or break covalent bonds in reactive groups usually indicates that receptors contain or are proteins. The use of different proteolytic enzymes, towards which the receptors for different viruses show different sensitivities, has been most rewarding, since cells remain viable after treatment at 37 °C with 1 % trypsin or chymotrypsin [748] or subtilisin [533]. Poliovirus receptors on intact or broken host cells [276, 528, 548, 748] or on solubilized host cell receptors [444] are sensitive to inactivation by trypsin. Polio receptors on intact cells are less sensitive to chymotrypsin than to trypsin [748] although receptors in cell lysates do not show differential sensitivity [528]. Conversely, cellular receptors [748] or solubilized erythrocyte receptors [528, 529] for coxsackie B3 virus and echovirus 7 are more sensitive to chymotrypsin than to trypsin. These differences in sensitivity are relative only. Cellular receptors for a simian enterovirus are also destroyed by trypsin and chymotrypsin [260]. The HeLa cell receptors for adenovirus type 2 are not inactivated by trypsin or chymotrypsin and attachment of virus is enhanced after this treatment; on the other hand, subtilisin destroys the receptors [533]. The receptors for adenovirus type 7 on erythrocyte membranes are, however, destroyed by trypsin or chymotrypsin [476], as are the erythrocyte receptors for reoviruses [387, 390]. HeLa cell receptors for human rhinoviruses are sensitive to trypsin and chymotrypsin [399], but the receptors for type 2 rhinovirus are far more sensitive than those for type 14 [LONBERG-HOLM, unpublished]. Since trypsin releases a large portion of the cell's carbohydrate [609], presumably in the form of glycopeptides, it is possible that the sensitivity of receptors to proteolytic enzymes results from the loss of, but not the destruction of, that portion which provides the specificity for virus-cell interaction, i. e. a polysaccharide.

Chemical reagents which primarily interact with protein components of the cell membrane have been applied to study virus receptors. The selective inhibition of erythrocyte receptors for adenovirus type 7 by 1-cyclohexyl-3-(2-morpholinoethyl)-carbodiimide metho-*p*-toluenesulfonate, a reagent modifying carboxyl groups, suggests that a protein carboxyl group is required in attachment [477]. Formaldehyde, which reacts with free amino groups, does not inactivate receptors for adenoviruses or influenza [476] or for a simian enterovirus [260]. On the other hand, formaldehyde and low molecular weight aldoses appear to react with and inactivate the hemagglutinin receptors for several hemagglutinating enteroviruses and reovirus type 2 [389]. Attachment of poliovirus to host cells does not involve the participation of sulfhydryl groups on the receptor [271] and reagents which block sulfhydryl groups do not block the receptors of host cells or erythrocytes for simian enterovirus 2 [260], the reoviruses [388], NDV [405], adenovirus 7 [476], vaccinia virus [6] or myxoviruses [6, 476]. In contrast, there is some evidence that the attachment of a coronavirus (avian infectious bronchitis virus) may give rise to a disulfide bond [405].

Virus receptors may also be inactivated by reagents which react primarily with carbohydrates. The most useful and widely tested reagent is the enzyme neuraminidase (receptor destroying enzyme, RDE, or sialidase) which releases terminal NANA from many glycoproteins and glycolipids. Myxoviruses and paramyxoviruses contain a virion neuraminidase. A more powerful and widely used enzyme is prepared from bacterial sources and can be obtained commercially. The cellular receptors for myxoviruses or paramyxoviruses are destroyed by neuraminidase [293]; excess of one virus can destroy host cell receptors for another virus and this provides the basis for one kind of interference among these viruses [23]. Attachment of an avian coronavirus was also inhibited by treating the host cells with neuraminidase [405]. Receptors of erythrocytes for polyoma virus are destroyed by neuraminidase [255, 462] but the host cell receptors are less sensitive [462]. The erythrocyte and host cell receptors for cardioviruses are sensitive to neuraminidase [361, 693]; but in the case of EMC virus, the receptors on L cells were less sensitive than those on HeLa cells [361]. The attachment of purified radioactive equine rhinovirus to HeLa cells is also prevented by treatment of the cells with neuraminidase [LONBERG-HOLM, unpublished] in contrast to the attachment of human rhinoviruses and poliovirus. Goose erythrocyte receptors for rabies virus [371], human erythrocyte receptors for adenovirus 7 [58] and for the 3 serotypes of reovirus [388] are all stable to neuraminidase while ox erythrocyte receptors for reovirus 3 (but not 2) are destroyed [228].

Glycosidases other than neuraminidase have been used infrequently but an $\alpha$-mannosidase from a marine gastropod has been shown to inactivate the receptor for Japanese encephalitis virus [745, 746]. Various other glycosidases have been tested, without effect, on the erythrocyte receptors for enteroviruses and reoviruses [387, 390].

Careful periodate oxidation of a glycoprotein receptor for influenza virus followed by borohydride reduction removed carbon atoms 7, or 8 and 9 from the NANA residues [657]. This removal blocked the binding of virus and, to a lesser extent, the release of the altered NANA residues by the virion neuraminidase. In addition to influenza virus receptors, the erythrocyte receptors for EMC [462] and for all reovirus types [228, 388] are destroyed by approx. $10^{-3}$ M periodate. It has been reported that $10^{-2}$ M periodate also inactivates the soluble erythrocyte receptor of the echovirus 7-coxsackie B virus family [528] and of adenovirus 7 [476] but these results are probably nonspecific. On the other hand, erythrocyte receptors for rabies virus [371] and a simian enterovirus [260], and host cell receptors for poliovirus [276, 444], are resistant to periodate.

Soluble receptor material for enteroviruses and arboviruses contain lipids and the inactivation of receptor activity by various lipolytic enzymes has been attempted. Lipase, a glycerol ester hydrolase, has been found to be inactive against receptors for a number of enterovirus [260, 276, 444, 548] but does show inhibitory activity against the receptors for western equine encephalitis virus [548]. Phospholipase C, but not phospholipase A, inactivated brain or erythrocyte receptors for Japanese encephalitis virus [461] while venom diesterase inactivated the soluble erythrocyte receptor for echovirus 7 [529]. Organic solvents also show varying abilities to inactivate virus receptors, for example, ether [276], chloroform [276, 444] and butanol [444] inactivate poliovirus receptors while butanol was used to solubilize active erythrocyte receptors for the hemagglutinating enteroviruses [529]. The lipids may therefore be necessary components in many virus receptors *in vitro,* but their function may only be structural.

## 3. Reversible Inhibition of Virions

Sera and other body fluids may contain substances which compete with receptors for the binding of viruses and may thus inhibit virus infectivity or hemagglutination; when such substances show a high enough binding constant they may appear to react irreversibly with virions. Serum inhibitors have been reviewed [5], and there is also a detailed review of myxovirus inhibitors [293] as well as a recently prepared review on hemagglutination inhibitors in general [291]. Several points should be established: (a) the multivalency of an inhibitor influences its potency; (b) specific soluble proteins may bind to virions, and (c) polyions, carbohydrates, and lipids may also interact with virions in a selective manner. In addition, it is difficult in practice to distinguish between those substances which interact with viruses and those which interact with receptors.

Glycoproteins containing sialic acid residues react reversibly with influenza and parainfluenza virions and are able to inhibit virus hemagglutination; these have been referred to as $\alpha$ or Francis inhibitors [293]. Similar material also inhibits hemagglutination of cardioviruses [426], reovirus [386], and polyomavirus [246]. A positive correlation exists between the molecular weight and sialic acid content of these substances and their ability to inhibit hemagglutination and this suggests that the sialic acid residues of an active material may present a complementary fit to the virion spikes [636]. A low molecular weight sialoglycoprotein (orosomucoid, molec. wt 41,000), which is by itself relatively inactive, was polymerized to a high molecular weight material which was a potent inhibitor of influenza virus hemagglutination and infectivity [454, 727, 728]. The filamentous structures of such polymers were found to be most potent [26]. The M and N blood group substances which form the normal HA receptors for influenza are cleaved from erythrocytes by trypsin and the resulting (monomeric) glycopeptides show little or no hemagglutination inhibition [615].

Protein inhibitors of influenza virus infectivity lacking neuraminic acid have also been found, suggesting that receptors other than sialoglycoproteins may also exist; these are called $\beta$ or Chu inhibitors and have also been reviewed [293]. A low molecular weight $\alpha_1$-globulin from human serum agglutinates adenovirus type 5 by attaching to the fiber antigens and thus inhibits viral hemagglutination [614]. A large unstable (19S) protein from human serum inhibits hemagglutination by echoviruses; it is neither an immunoglobulin nor a lipoprotein [66]. Serum $\beta$-lipoprotein has also been found to inhibit hemagglutination of reovirus [428]; and a lipoprotein released into serum-free tissue culture fluids also inhibits reovirus and enterovirus hemagglutination but this substance has properties which distinguish it from cellular receptors for the enteroviruses [597]. Some of the 'nonspecific' protein inhibitors may be related to virus receptors on host cells; if so, antisera to the inhibitors might be used to establish such a relationship [662].

Polyanions can interact with virions and compete with cellular receptors. It was first recognized that sulfated polysaccharides from agar inhibited the development of the plaques of certain viruses [666]. Attachment, plaque size and the efficiency of plaque formation of influenza virus [664] or of herpes simplex virus [665] were inhibited by acid polysaccharides isolated from agar and the effects were overcome by the polycation DEAE-dextran which probably binds the anionic polymers. An arbovirus of group B (dengue) was shown to react with sulfated polysaccharides from agar which prevented its attachment to cells; the reversible complex formed with the virus was isolated by sedimentation and then shown to be dissociable [601]. Dextran sulfate, a synthetic acid polysaccharide, has also been shown to prevent the attachment of Japanese encephalitis virus [510] and herpesvirus [665] to host cells. The attachment of respiratory syncytial virus [489] and FMD virus [33], and of some variants of poliovirus [32] and of

EMC virus [396] are also inhibited by dextran sulfate, and other negatively charged polymers inhibit the attachment of fowl plague virus [10]. Heparin sulfate, an acid polysaccharide of animal origin, also prevents attachment of herpesvirus to host cells [468, 665] and has been employed to dissociate infective virus from virus-receptor complexes early after infection [267]. The relationship of the pH to the relative effect of the polyanion on the adsorption of EMC to cells [396] and the effect of pH on the adsorption of poliovirus to immobilized cationic and anionic exchangers [34] are both sufficiently complex to defy any simple model for the effects of dextran sulfate. Dextran sulfate or other polyanions may also interact with the cell [666]. Inorganic polyanions such as silicotungstates or borotungstates have been reported to have antiviral activity in tissue culture systems [552], and sodium hexametaphosphate interferes with hemagglutination by EMC virus [215].

Carbohydrates can interact with virions in competition with cellular receptors. The interaction of sialomucoproteins with influenza virus is a good example. Monosaccharides have been noted to interfere with the hemagglutination of various enteroviruses [387] and with EMC virus [368]. In the first case, however, there is some indication that the carbohydrates interact primarily with the receptors. N-acetyl-D-glucosamine has been claimed to interact with the reovirus virion to prevent hemagglutination, but 10% w/v of the sugar had to be used [216]. A stronger case for the interaction of carbohydrate with virions has been made for Japanese encephalitis virus. D-Mannose and related sugars inhibited hemagglutination [499] and $\alpha$-mannosidase destroyed the activity of a purified cellular receptor [745]. It was also found that the polysaccharide mannan constituted a powerful inhibitor of hemagglutination [285, 498].

Lipids play a structural role in the cell membrane but whether they play more than this role in cellular receptors is not known. Lipids constitute about one third by weight of the partially purified soluble cellular-receptor preparations for enteroviruses [444, 529]. Materials with the properties of lipoproteins inhibit hemagglutination by enteroviruses and reoviruses [428, 597] as already discussed. These activities appear to be related to the effect of lipids, including phospholipids and sphingomyelins, on enterovirus or reovirus hemagglutination [598], and suggest that lipids interact with the virions. The lipids used, however, were not pure and their activity might have been due to contaminating glycolipids. In the case of arboviruses, mixtures of lipids or phospholipids can also inhibit hemagglutination [211, 230] as already discussed in section II.C.

## 4. Reversible Inhibition of Receptors

Virus components can block virus attachment as exemplified by adenovirus fiber [533]. Antibodies can also react with receptors (see below). Less specific interactions may also lead to the inhibition of virus attachment. The polycation, protamine sulfate, may compete with virus for cellular receptors leading to inhibition of infectivity of respiratory syncytial virus [489] and mengovirus [108]. At optimal concentrations, however, protamine may enhance infectivity, presumably by interacting with sulfated polysaccharides or other polyanions in the overlay.

Streptomycin, a basic antibiotic, and protamine have both been used for the elution of arbovirus from its cellular receptors and this may be caused by competition for the receptors [461]. DEAE-dextran inhibits attachment of Japanese encephalitis virus to host cells [510]; however, DEAE-dextran is used more widely as an enhancer (see below).

Lerner *et al.* [387] discovered that various simple sugars, at high concentrations, inhibited or altered the normal patterns of hemagglutination of enteroviruses and reoviruses. A free carbonyl group was required for activity; glyceraldehyde was more active than hexoses or pentoses but formaldehyde had the highest activity [389]. The aldehydes bound to the erythrocytes in preference to the virions, leading to the conclusion that inhibition of hemagglutination was caused by the reaction of the carbonyl groups of the carbohydrates with the cellular receptors. This hypothesis may be irrelevant for cellular infection if highly purified enteroviruses lack carbohydrate (see discussion above). It is, however, hypothetically conceivable that a carbohydrate or a carbohydrate-containing molecule might serve as a cofactor for attachment.

## 5. Virus-Specific Antibodies and Attachment

The ability of neutralizing antibodies to block hemagglutination has been employed in diagnostic virology [340] and will not be reviewed. It does, however, establish that antibodies to virus can interfere with attachment to receptors. The virus-specific antibodies which affect attachment can be distinguished from others which also cause neutralization [421]. Neutralized NDV virus can still attach to host cells [577] and the attachment of neutralized poliovirus may even be enhanced by IgG under some conditions [421]. It should, however, be emphasized that antibodies may aggregate virions [543] and that these aggregates may sediment together with cells and thus appear to be cell-associated.

When the rate of attachment is enhanced by antibody and when neutralization is inefficient, there can be an enhancement of virus infectivity by antiviral serum. This was observed when avian antibodies to rabbitpox virus or to a group B arbovirus were tested in an avian cell system [257]. An alternative explanation for this effect may be that antiviral antibodies stabilize the virions while they accomplish successful infection; *E. coli* galactosidase shows increased activity if incubated with antibodies [573a]. Additional evidence for a distinction between neutralization and inhibition of attachment comes from the observation that treatment of FMD virions with trypsin impairs their ability to attach to host cells but does not eliminate antigenic determinants which elicit neutralizing antibodies [576]. The possible molecular aspects of neutralization will be discussed in section V.

### 6. Host-Specific Antibodies and Attachment

Anticellular antibodies can also inhibit attachment of animal viruses to host cells. Anticellular sera were readily shown to inhibit the infective process of poliovirus [276, 546], some types of coxsackievirus A [239], echoviruses [239, 546] and adenoviruses [239]. Furthermore, antihuman sera were effective in monkey cells, which implied that a surface component of the monkey cell shares antigenicity with its human counterpart [239, 276]. Antibodies prepared against a semipurified receptor for influenza virus were able to block infectivity [662]. In the case of enveloped viruses, however, the anticellular sera might react with a host cell component, presumably a glycolipid, in the virion envelope and this may be a basis for antiviral activity [363, see also references therein]. Although anticellular antibodies may inhibit steps subsequent to attachment (sect. V), their primary action is probably to block virus receptors. It has been clearly demonstrated that the attachments of polioviruses, echovirus 6 and coxsackieviruses types A13, B1 and B3 to HeLa cells are inhibited by anticellular serum [14]. Antibodies which inhibited attachment of poliovirus could be adsorbed with chymotrypsin-treated cells or cells whose coxsackie B virus receptors were saturated with excess virions. Heated cells, which lack poliovirus receptors, have also been used to produce serum specific for the receptors for coxsackie B viruses [466].

## 7. Enhancement of Attachment

Reagents which enhance the rate of attachment may provide evidence for the nature of the virus-receptor bond. Treatment of intact HeLa or KB cells with trypsin or chymotrypsin enhances their ability to attach adenovirus type 2 [533] and under certain conditions may also enhance attachment of poliovirus [748]. Pancreatin treatment of avian cells also enhances the uptake of a coronavirus [405]. Trypsin treatment also causes the randomly distributed morphological units of the membranes of erythrocytes [431, 630] and certain receptor units of growing cells [483] to aggregate or group into patches. This rearrangement of the cell surface is probably due to a reduction in the repulsive forces between the membrane subunits so they can undergo lateral aggregation. The enhancement of virus attachment may consequently be due to an increased density of receptor subunits in the patches, allowing for the multivalent bonding of virus. Indeed, attachment of influenza virus to trypsin-treated erythrocytes has been shown to occur on the patches [431]. In cases where trypsin destroys the viral receptors, there of course may be no enhancement. In addition, enzymatic treatment of the cell may alter its net charge so as to facilitate the attachment of virions. Although this has not been established, it has been suggested to explain the enhancing effect of neuraminidase treatment on attachment of adenovirus to host cells [58] and the enhancing effect of neuraminidase and phospholipase C treatment of erythrocytes on the hemagglutination of various leukoviruses [661]. In these cases it is also possible that the rearrangement of receptors into patches is the basis for enhanced attachment. Furthermore, fucosidase treatment of neuraminidase-treated cells produces an additional enhancement of adenovirus attachment [58] suggesting again that decreased hydrophilicity of the membrane glycoproteins may permit rearrangement into patches. Other explanations have been suggested, such as a decrease in the rigidity of the membrane or a removal of steric impediments to virus-receptor interaction [58].

Polycations have also been noted to enhance the infectivity of many viruses. DEAE-dextran, present during virus adsorption, increases the efficiency of plaque formation by rabies virus [329], respiratory syncytial virus [489], and rubella virus [687], and of focus-formation by a sarcoma virus [694]. DEAE-dextran has been shown to enhance the rate of attachment of fowl plague virus to chick cells [10]. The enhancement here may result from interaction of the polycation with the receptors rather than the virions, since DEAE-dextran has profound effects on the biochemistry of cells and hence must interact with cells [65]. It is possible that DEAE-dextran alters the spatial arrangement of subunits on the cell surface, or that it decreases the net charge of the surface, or that it serves as a ligand between virus and cell; it is also well known that DEAE-cellulose can be used as a selective adsorbent for some virions [34]. DEAE-dextran may also bind to polyanions which otherwise would have an inhibitory effect on the attachment (see above).

There appear to be naturally occurring enhancers for virus attachment. An antigen produced by one avian oncornavirus (RAV-1) enhances the ability of RSV(RAV-2) to attach to susceptible cells; this enhancement is achieved by incubating RSV(RAV-2) virions with extracts of RAV-1-infected cells and may be due to the addition of RAV-1 antigenic determinants to the RAV-2 virion envelopes [248]. The rate of attachment of radioactive RAV-2 virions was increased after this modification. An enhancer for the attachment of a plaque variant of echovirus type 6 to nonpermissive cells has been detected in extracts of permissive cells [556]. Recent evidence indicates that the enhancer consists of fragments of the membrane of the permissive human cells and that *in vitro* incubation with the virus results in formation of virus-receptor complexes which are themselves infective in the nonpermissive simian cells [555]. In these cases, as with DEAE-dextran, virus attachment was enhanced by the addition of a third macromolecular component or cofactor. Such a third component may interact primarily either with the virion or with the cell but ultimately helps to form the bond between virus and cell.

## *C. The Mechanism of Attachment*

### 1. Reversibility

After virus has become bound to cells sufficiently firmly to become 'cell-associated' by an experimental criterion (for example after washing the cells with virus-free medium), it may pass through several stages leading from a loose to a tight bond with the cell. These stages are detected and defined by changes in the ability to recover the virus from the cell. The attachment may be said to be 'reversible' if infective virus can be dissociated under physiological conditions, such as incubation in virus-free medium. Infective influenza virus may be eluted from cells by repeated washing [4] or by incubation [9] but this usually reflects the ability of the virion neuraminidase to hydrolyze the sialic acid of the receptors. EMC virus contains no neuraminidase but it also can elute after attachment in the cold and subsequent incubation at physiological temperature, as determined with erythrocytes [375, 747] or HeLa cells [LONBERG-HOLM, unpublished]. Polyomavirus, which also contains no receptor destroying enzyme, behaves in the same way [119]. Poliovirus-1 and echovirus-7 have also been reported to form a reversible attachment with HeLa cells at 0 °C so that dilution of virus-cell complexes results in detectable elution of infectivity [528]. The attachment of other viruses appears to require divalent cations and in these cases EDTA can elute virus which has attached in the cold whereas virus attached and incubated at physiological temperatures is not elutable (see below). In conclusion, there is growing evidence that the attachment of virus may involve a sequence of events leading from loose to tight binding and progression through this sequence may depend upon temperature.

## 2. The Rate of Attachment

The rate of attachment of virus obviously depends upon the concentration of receptors, i. e. upon the number of cells per ml. If less than saturating levels of virus are employed, and if the cell concentration remains constant, the adsorption of virus will follow approximately first order kinetics; the rate is thus independent of the virus concentration and the apparent rate constant can be expressed in the units *ml per minute per cell.* Work on the attachment of phage particles to bacteria in the early 1930's attempted to relate the observed rate of attachment to the maximum rate of attachment which was calculated from equations for the frequency with which virions collide with the cell surface by Brownian motion, and with the assumption that all collisions lead to attachment [594]. ALLISON and VALENTINE [9, 10, 688] later attempted to define the limitations under which such equations could be applied to animal virus-host cell systems and compared the adsorption curves obtained from the integrated theoretical equations with experimental results obtained with virus and cells or virus and nonbiological surfaces. The attachment of vaccinia or fowl plague virus to nonbiological surfaces approached the theoretical maximum rate but the attachment to cells occurred at a rate $^1/_2$ to $^1/_3$ of the maximum. To correct for this discrepancy, an efficiency factor was arbitrarily introduced; this factor represented the fraction of the collisions which lead to permanent attachment ($^1/_2$ to $^1/_3$). Unfortunately, this factor has since been shown to have no physical meaning since one of the basic initial assumptions of the equation for the collision frequency was that the concentration of free virus is zero at the cell surface, i. e. all virus becomes attached when in contact with the cell [359, 500]. In fact, it has been calculated that the collision frequency must be $10^4$ greater than the frequency of attachment if the rate of attachment is about half the theoretical maximum as calculated by the equations of VALENTINE and ALLISON [359]. As a metaphor, a virion must knock 10,000 times before the door is answered.

Thus the efficiency of attachment of virions which collide with a cell is exceedingly low, and not close to one as has often been assumed. The low efficiency of attachment agrees with our observations that only a small portion of the cell's surface is covered with virions at the saturation of the cell receptors (see above). Even with such low efficiency for attachment, the rate of attachment is within a factor of 3 of the maximum theoretical rate for vaccinia and fowl plague virus, and more important, the effect of temperature on the attachment of these viruses is relatively slight as is predicted from the calculations of the maximal rate as based upon collision frequencies [9]. Unfortunately, however, the attachment of many other viruses to cells does not show these properties.

## 3. Factors Affecting the Rate of Attachment

The composition of the medium may affect the rate of virus attachment. In some cases virus-receptor interaction is sensitive to changes in pH. Echovirus type 7 attachment to erythrocyte membranes has a sharp optimum between pH 5 and 6 [528], whereas coxsackievirus B4 attaches to HeLa cells with a sharp optimum between pH 3.0 and 3.5 [125]. The optima for several other enteroviruses [125], a cardiovirus [244] and NDV, a paramyxovirus [581a], all show broad peaks close to or encompassing neutral pH, but each has its own characteristic range. It is interesting to note that several coxsackie B viruses, which all presumably share receptors [122], have distinctly different pH optima for attachment [125]. It is difficult to attribute the effect of pH simply to the dissociation of an ionized group on the receptor. Adenovirus 7 attachment to erythrocyte membranes shows a broad pH optimum between 5.5 and 8.7 and in this system some chemical evidence has been provided to suggest that the bond involves a positively charged arginine on the virion and an ionized carbonyl group on the receptor [477]. Poliovirus also shows a broad optimum for attachment between pH 4.5 and 8.5 [125]. In the case of poliovirus, however, it has been shown that the virions undergo profound conformational alterations in the range 4.0–7.0 and that these alterations produce major changes in the net charge of the virions [424]. The binding of various strains of poliovirus to polyions also varies with pH but in a manner which defies a simple explanation based upon net charge alone [34]. We are left with the conclusion that virus attachment may involve specific electrostatic interactions but that net charge alone is an insufficient indicator of whether or not binding can occur.

The ionic strength of the medium also appears to influence the rate of attachment. NDV [581a] and type 1 poliovirus [276] attach poorly in isotonic sucrose but the rate is returned to normal when NaCl is added. The attachment of poliovirus is not appreciably inhibited by 1 M NaCl [677] but on the other hand, very high concentrations of NaCl (5 M) can elute virus from the receptors [271]. The optimum attachment of adenovirus type 7 also requires 0.1 M salt [477] as does the attachment of phage R17 [145] but the attachment of coxsackie B3 may be enhanced in hypotonic medium [124].

The presence of divalent cations may also influence the rate of attachment. It was reported that poliovirus attachment was enhanced by $Mg^{++}$ or $Ca^{++}$ but the monkey cells employed had been trypsinized [19] and such treatment alters the number and characteristics of the cellular receptors. Other work with nontrypsinized cells indicates that divalent cations are not needed for the attachment of poliovirus [276, 677]. Divalent cations enhance, and EDTA inhibits, the attachment of coxsackievirus A9 [443], human rhinovirus type 2 but not 14 [399; Noble-Harvey and Lonberg-Holm, in preparation] and FMDV [68] to host cells. In each case where divalent cations have been found to be required in attachment, it has also been found that incubation of the virus-receptor complex with EDTA can produce dissociation of infective virus, provided that the virus has attached in the cold or has been incubated with cells for only a short time at physiological temperature.

## 4. The Effect of Temperature

Some viruses, notably viruses which recognize sialic acid-containing receptors, appear to attach rapidly to host cells and this attachment is not temperature sensitive. This group includes the myxoviruses [9, 311], cardioviruses [286, 324, 368, 414], polyomavirus [119], and equine rhinovirus [399].

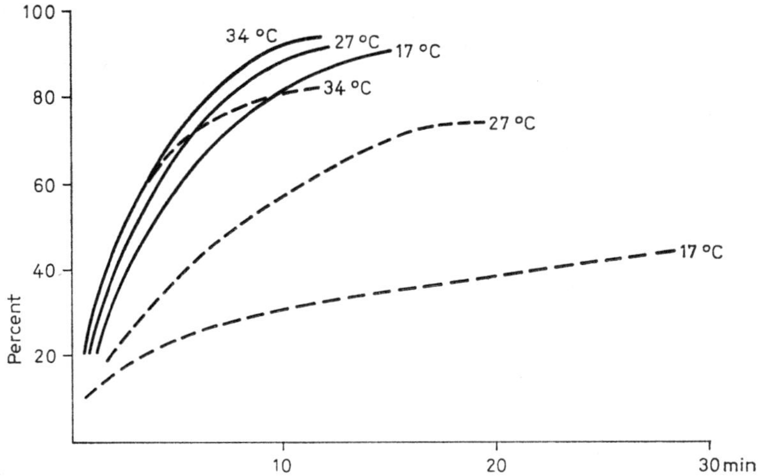


*Fig. 5.* The effect of temperature on attachment of rhinovirus type 2 virions to HeLa cells. *Dashed lines* indicate experimental results for attachment of $^3$H-virions to $10^7$ cells per ml [399]. The *solid lines* were calculated by equation 3 of ALLISON and VALENTINE [9], with an efficiency factor (∅) of 0.17 and with corrections for changes in viscosity and the diffusion coefficient of virions at the indicated temperatures (computations were made by M. UHLEN).

The attachment of other viruses is less efficient and shows sensitivity to reduced temperature. Figure 5 illustrates the experimental attachment of human rhinovirus type 2 to suspended HeLa cells at 3 temperatures (dashed lines), and the attachment calculated from the equation of ALLISON and VALENTINE [9] (solid lines) in which we have used an 'efficiency' factor of 0.17 (the initial attachment of rhinovirus at 34 °C is 0.17 times the maximal rate limited only by diffusion). Although this equation is generally invalid, it does permit the valid correction for the effect of temperature on the maximal collision frequency of virus and cell. The observed rate of attachment of virus decreases greatly as the temperature is reduced; this decrease is much greater than would be expected from the effect of temperature on the maximal frequency of collision. With polioviruses 1 and 2 and rhinoviruses 2 and 14, the rate of attachment to suspension cultures of HeLa cells shows an increase by a factor of 2–3 per 10 °C increase in temperature in the range of 10–37 °C [399; LONBERG-HOLM, unpublished]. This would imply an activation energy of up to 20 kcal if a single chemical bond were involved in the attachment reaction. It seems, however, unlikely that attachment involves the formation of a covalent bond, since the infectivity of attached virus is initially unaltered. The attachment of coxsackie B viruses [CROWELL, private commun.], of adenovirus type 2 [525] and an arbovirus [553] has also been found to be strongly temperature-dependent.

It has, however, been reported that the rate of attachment of poliovirus to *in vitro* receptors is reduced only $1/2$ by cooling from 25 to 0 °C but it should be noted that this did not involve intact cells [677]. Other reports that the attachment of poliovirus type 1 to suspended HeLa cells is relatively insensitive to temperature are based on the use of trypsinized cells [19, 276]. The attachment of poliovirus type 1 to monolayers of HeLa cells showed a 10-fold reduction in rate at 1 °C relative to 37 °C [276]. Note, however, that the effect of temperature on the attachment of poliovirus, or other viruses discussed above, to nontransformed cells needs to be investigated further.

An effect of temperature on the attachment of virus might reflect a cellular metabolic step required for the formation of a virus-cell complex. However, it has not yet been shown, despite several attempts, that the attachment of virus is sensitive to metabolic poisons. The effect of temperature might also reflect a requirement for conformational alterations in either virions or cell membrane components during attachment.

## 5. A Model for Virus Attachment

We would like here to propose another hypothetical basis for the effect of temperature. Virus may first interact reversibly with one or a few specific receptor subunits on the surface of the cell with a binding constant too low to permit detectable attachment after the cells have been sedimented and/or washed to remove unattached virus. Additional receptor subunits are able to diffuse in the plane of the cell membrane to the initial unstable virus-receptor complex and these also interact with the virus to produce multivalent binding which becomes firm enough to allow the detection of virus attachment. The diffusion of the receptor subunits in the plane of the cell membrane must be greatly dependent upon the temperature since the viscosity of the lipid-containing membrane varies greatly over the range 0–37 °C. Since each step in the establishment of such multivalent binding is reversible, virus attached to cells may sometimes be dissociated upon repeated washings or upon incubation. Ultimately, the bond will be so strong that the attachment will appear to be irreversible, unless reagents are employed which alter the dissociation constants.

If divalent cations participate in the binding of each receptor subunit to virus, their removal by EDTA would drive the reaction in the direction of dissociation, but virus which had become firmly bound to the cell by many receptors, or which had penetrated the cell, might resist dissociation. It is also possible to view the effect of divalent cations as permitting the proper conformation or juxtaposition of receptor units on the cell surface.

The multivalent model explains why there are more receptors for adenovirus fiber proteins than receptor sites for intact virions, providing that it is assumed that the binding constant for the smaller fiber protein to a single receptor is greater than that of the large virions. The dependency of the overall dissociation constant upon the multivalency of the virus-cell bond is also similar to the effect of polymerization on the effectiveness of hemagglutination inhibitors described above. It can also be predicted that isolated receptors may show only relatively weak binding for virus and this may provide an obstacle for their eventual purification and identification.

The reason why certain viruses, which recognize sialic acid-containing receptors, are bound effectively at low temperatures may be the abundant distribution of their receptors on the cell surface, i. e. the establishment of multivalent binding may not depend upon the diffusion of receptors in the plasma membrane if these are already located close to each other. As described above, certain viruses show more rapid attachment after cells have been subjected to proteolytic digestion. Proteolytic digestion produces a rearrangement of the membrane subunits into 'patches' on the cell surface. Receptor units located in patches might then be close enough so that they could participate in multivalent binding to a virion without need for diffusion.

In conclusion, the attachment of virus to cellular receptors involves many specific interactions which at the present time cannot be reduced to complete analysis. These systems are probably even more complex than the binding of antibody to antigen.

# IV. Virus-Erythrocyte Systems

## *A. Specificity and Identification of Receptors*

The observation that fowl erythrocytes were agglutinated by influenza viruses [265, 440] led to the discovery of the first viral enzyme, neuraminidase, which is unique to myxo- and paramyxoviruses [231, 232, 266]. The subsequent delineation of hemagglutination for several other groups of viruses has provided model systems for the study of their early interaction with cells. The vast amount of data accumulated reveals that hemagglutination differs between different virus groups and also between strains. This establishes the complexity of the patterns of early virus-cell interaction. At the same time these studies have also offered insight into the structure and composition of the erythrocyte membrane and the molecules recognized by viruses as receptors.

We will review, as examples, the identification of receptors for one group of viruses with helical symmetry, the orthomyxoviruses, and two different viruses with cubical symmetry, the adenoviruses and the picornaviruses. The parameters of hemagglutination for these viruses are tabulated in table II. Most of the available data in this area of research have recently been reviewed by Howe and Lee [291], and we will here only emphasize that many viruses have receptors on the erythrocyte membrane.

*Myxoviruses.* The reaction of orthomyxoviruses with red cells involves adsorption at 4 °C and elution at 37 °C. The viral hemagglutinin [719, 720] and a neuraminidase enzyme which effects elution [231, 740] both reside as separate spikes on the virion envelope and are independently coded by the virus genome. Biologically active orthomyxovirus neuraminidase and hemagglutinin have been dissociated from virions [380, 718–720] and resolved, and their polypeptides have been identified [350, 603, 628,

*Table II.* Parameters of hemagglutination for different animal viruses

| Virus group | Subgroups | Erythrocyte species | Temperature for detection °C | Receptor alteration | Receptor chemistry | Receptor viral site | Receptor inhibitors |
|---|---|---|---|---|---|---|---|
| Myxo | A, B, C | fowl, various mammalian | 4 eluted at 37 | NANA released by virus; sialopeptide by trypsin | sialoglycoprotein | envelope spikes | $\alpha$-sialoprotein, $\beta$-serum, horse serum |
| Paramyxo | NDV, mumps, parainfluenza 1–5 | fowl, various | 4 eluted at 37 | NANA released by virus; sialopeptide by trypsin | sialoglycoprotein | envelope spikes | ovine $\alpha_1$ glycoprotein (NDV) |
| Pseudomyxo | | primate only | 37 | inactivated by trypsin and formalin | ? protein | envelope and subunits derived by Tween-ether | human sera |
| Adeno | I | monkey (complete) | 20 or 37 | inactivated by proteases (RDE for type 9) | ? protein | capsid, bivalent subunits, dodecons | serum |
| | II | rat (complete) | | | | | |
| | III | rat, human (partial) | | | | | |
| Entero | echo, coxsackie | human | 4 or 37 | chymotrypsin and formalin inactivates | protein or lipoprotein | capsid | serum, tissue culture media |

Abbreviations: NANA = N-acetylneuraminic acid; RDE = receptor destroying enzyme from *V. cholerae*. Modified from Howe and Lee [291].

638]. The neuraminidase has a molecular weight of 40,000 while there are two hemagglutinin polypeptides, HA1 and HA2, which have molecular weights of 70,000 and 50,000, respectively.

It should be stressed that hemagglutination activity is dependent on the state of aggregation of the hemagglutinin polypeptides, and hemagglutination requires at least a divalent hemagglutinin unit in order to bind erythrocytes to each other. Antibodies against the viral hemagglutinin or other substances which react with it can inhibit the attachment of virus to the erythrocyte membrane, and this provides the basis for hemagglutination inhibition. The receptor on the red cell is destroyed by viral neuraminidase when the virus-cell complex is warmed to 37 °C [266] and so it was only after it was shown that a soluble receptor could be extracted from erythrocyte stroma with 50 % hot phenol [336] that the exact chemistry of the receptor could be delineated. Virus receptor substance (VRS) extracted by this procedure is a sialoglycoprotein, apparently of 55,000 molecular weight [289, 291, 336, 635]. This molecular weight may, however, be questionable since glycoproteins behave anomalously in SDS polyacrylamide gel electrophoresis [65a]. Recently it was shown that the yield of this glycoprotein could be enhanced if the membranes were extracted with lithium diiodosalicylate prior to the phenol extraction [429]; other extraction methods include the use of pyridine [45] and butanol [289, 409] but these are not as selective. VRS contains most of the NANA of the erythrocyte membrane [114, 144, 290] and also the M and N blood group determinants [289, 336, 635].

Proteolytic enzymes like trypsin inactivate the receptor activity of isolated VRS, probably as a result of the release of the glycosidic entity as small sialopeptides [114]. NANA appears to be terminally bound to the short oligosaccharide side chains which are in turn bound to the linear polypeptide through 0-glycosidic linkages of N-acetylgalactosamine to either serine or threonine, or through N-glycosidic linkages of N-acetylglucosamine to aspartic acid residues [737]. Recent evidence suggests that the antigenic determinants are contained in alkali-stable trisaccharides obtained by borohydride degradation of the VRS from which NANA has previously been removed by mild acid hydrolysis [178]. NANA at the terminal nonreducing position is $\alpha$-ketosidically linked to N-acetylgalactosamine and this bond is cleaved by the viral neuraminidase so that free NANA is liberated [84, 231, 232]. It is of considerable interest that low molecular weight oligosaccharides used as substrate are cleaved with different specificities by a bacterial enzyme, the classical receptor destroying enzyme (RDE) from *V. cholerae,* and by influenza neuraminidase. The bacterial enzyme hydrolyzes both a 2–6' and a 2–3' ketosidic linkage to liberate NANA, whereas the viral enzyme only cleaves a 2–3' bond, although it binds the 2–6' compounds but does not release NANA [163].

Different myxovirus strains show differing affinity for the receptors at the cell surface. In general, the completeness of elution from erythrocytes appears to be inversely proportional to the sensitivity to glycoprotein hemagglutinin inhibitors [96, 97]. The strong interaction of so-called + strains of A2 influenza virus with both red cells and the gamma inhibitor of horse serum is probably due to a failure of the viral neuraminidase to cleave all of the receptors, but it may also be caused by an ability of the + strains to recognize additional receptors [96, 97].

The carbohydrate structures of blood group antigens are also added to glycosphingolipids, presumably by the transferase enzymes used for the glycoproteins [223a, 643a], and synthetic liposomes, to which have been added crude mixtures of gangliosides, have been shown to attach paramyxoviruses *in vitro* [HAYWOOD, personal commun.].

It appears established that the hemagglutinin and erythrocyte receptor of the paramyxoviruses are similar to those of the orthomyxovirus [292]. Paramyxoviruses also carry a neuraminidase although they elute from cells with lower efficiency [592, 593] and also generally induce hemolysis, as first described for mumps virus [99, 460]. Hemolysis is demonstrated best at 37 °C and involves a mechanism which may be related to the process which results in the fusion of the erythrocyte membrane and the viral envelope. Both the hemagglutinin and the hemolysis are trypsin-sensitive but it appears evident that they may reside in different molecules since hemolysis can only be demonstrated with infective virions or aggregated subunits [287].

The pseudomyxoviruses (measles, canine distemper and rhinderpest) resemble in many respects the paramyxoviruses since a hemagglutinating member, measles virus, has both hemagglutinin and hemolysin; on the other hand they all lack neuraminidase [714]. It appears that the envelope of measles virus is necessary both for hemagglutination and for hemolysis but the biochemistry of the receptor moiety on the erythrocyte is still unknown [495]; it has been suggested that the erythrocyte receptor differs from those of ortho- and paramyxoviruses [518, 684]. This suggestion is based partly on the finding that erythrocytes treated with formaldehyde showed no receptor for measles although the receptors for ortho- and paramyxoviruses were intact [518].

*Adenoviruses.* All 32 serotypes of human adenoviruses have been shown to hemagglutinate [29, 493, 570–572]. A classification based on hemagglutination was proposed by ROSEN [571] and is shown in table II. Subgroup I agglutinates monkey, preferably rhesus and grivet, erythrocytes with a complete pattern while subgroup II agglutinates rat erythrocytes, also with a complete pattern. Finally, subgroup III viruses agglutinate rat cells with a poorly defined or partial pattern. Subdivisions within these groups have also been suggested [492, 493].

It has been established that the vertex region of the virion is primarily involved in the interaction between viruses and red cells and that type-specific determinant of the fiber carries this capacity [491, 496, 517, 521, 689]. Since the intact adenovirus carries 12 vertex projections it can establish a bridge between two erythrocytes and cause hemagglutination. Intact penton and/or fiber units are incomplete hemagglutinins because they can only establish a monovalent link with a cell. They can be made into complete hemagglutinins if antibodies are added to cross-link them [491, 493, 517] or if they aggregate naturally into multimeric units [490, 700, 701]. Since the distal part of the fiber is recognized by the receptor, and this contains type-specific antigenic determinants, heterotypic antibodies which react with the base of the fiber can be used to generate complete hemagglutinins [497, 517] and this has been called *hemagglutination enhancement* [517]. The fibers of different adenovirus types differ in length and the short fibers of subgroup I and II cannot cause hemagglutination in the presence of heterotypic antisera because they lack a necessary antigenic determinant at the proximal end of the fiber [497]. A complete subviral hemagglutinin of adenoviruses within subgroup I and subgroup II, the 'dodecon', is composed of 12 penton units arranged as an icosahedron with fibers protruding [490, 493].

Adenovirus receptors of the erythrocyte membrane have not been defined biochemically but they have been solubilized and partially purified from monkey red cells [476]. This receptor material does not interact with myxoviruses and is not sensitive to neuraminidase and is thus distinct from the VRS. Phenol, butanol and potassium iodosalicylate treatment of membrane preparations render them incapable of inhibiting the hemagglutination of adenovirus type 7 and the activity was also lost after treatment with proteolytic enzymes [476]. Adenovirus type 9 within subgroup II shows unique properties since receptors can be removed by RDE and virus can be eluted from cells with this enzyme [698]. None of the adenoviruses carries enzymatic activity which digests receptors and thus permits elution.

*Picornaviruses*. Among picornaviruses, the echovirus group, the coxsackieviruses and a variety of unclassified enteroviruses show hemagglutinating properties with human or primate erythrocytes [226, 522] while cardioviruses (EMC, ME, mengovirus, etc.) agglutinate a wider variety of erythrocytes [155, 215]. Some rhinoviruses also agglutinate sheep erythrocytes [644]. Although hemagglutination can often be detected within a serotype of picornavirus, all individual strains of this type may not hemagglutinate, in which case antibodies against a nonhemagglutinating subtype inhibit agglutination by an indicator strain. Certain serotypes attach to the red cells at 4 °C but elute at 37 °C; this elution is nonenzymatic, and erythrocytes are fully agglutinable after virus elution, in contrast to the orthomyxoviruses [339, 537].

A soluble component obtained after butanol extraction of red cell membranes in high salt concentration was found to have receptor activity and was insensitive to RDE [529]. This component contained protein, lipids and carbohydrates, but only low levels of neuraminic acid, and it inactivated hemagglutinating enteroviruses 1,000 times more efficiently than myxoviruses.

The erythrocyte receptors for picornaviruses are not all similar to each other, however. The receptors for cardiovirus hemagglutination [155] and for hemagglutination of some rhinoviruses [644] are destroyed by treatment with RDE.

In conclusion, the receptor material for enteroviruses on erythrocytes may be a lipoprotein and may differ from that on host cells [529]. Receptor material from host cells also appears to be lipoprotein as do inhibitors derived from tissue culture fluids, normal and infected cells, and different animal sera. Receptors for enteroviruses are also distinct from those for nonenterovirus picornaviruses, from those for myxoviruses and probably also from those for adenoviruses.

### *B. Mechanism of Interaction*

*Myxoviruses*. Adsorption of the myxoviruses appears to be independent of temperature, occurring as well at 4 as at 37 °C. The primary step has been regarded as an electrostatic interaction [683]. The study of various types of glycoproteins as receptor analogues has defined the molecular basis of a receptor and also the requirements for the action of neuraminidase. The activity of these naturally occurring glycoproteins is dependent

*Table III.* Equilibrium parameters for virus-inhibitor and antigen-antibody interaction

| System | Inhibitor | Molecular weight | Equilibrium constant | Inhibitory concentration, μmol[a] | |
|---|---|---|---|---|---|
| | | | | biased assay system | unbiased assay system |
| Myxovirus[b] and inhibitor | OSM[c] | $1 \times 10^{6}$ | $1 \times 10^{-8}$ | 0.0005 | 0.05 |
| | trypsin-treated OSM | $5 \times 10^{4}$ | $4 \times 10^{-6}$ | 9.5 | 94.5 |
| Antibody[d] and antigen | ($\alpha$-1,6-glucose)$_n$ | $7.5 \times 10^{4}$ | $1–2 \times 10^{-8}$ | 0.07 | NT[e] |
| | glucose | $1.8 \times 10^{2}$ | $2 \times 10^{-5}$ | 250 | NT |

[a] The concentration given caused 50% inhibition of hemagglutination or immunoprecipitation, respectively. In the biased system the inhibitor was allowed to react with virus or antigen prior to addition of assay system. In the unbiased system all three reactants were added concurrently.

[b] Data from FAZEKAS DE ST. GROTH and GOTTSCHALK [196].

[c] OSM = Ovine submaxillary mucoprotein.

[d] Data from KABAT [326].

[e] NT = Not tested.

on their molecular weight [636, 728] and this is compared in table III with the dependency of the association constant between antigen and antibody as a function of the size of the antigen. Polymerization of hemagglutination inhibitors also greatly enhances their binding to virions as discussed in section III. The oligosaccharide side chains of the membrane proteins may not be the only recognition units since treatment of erythrocytes with trypsin removes most of their NANA but does not entirely destroy their ability to be agglutinated by influenza virus [738]. Treatment with neuraminidase drastically reduces the net negative charge of the erythrocyte as detected by its electrophoretic mobility [114, 250, 645] and most of the net charge of the untreated cell can thus be ascribed to ionized free carboxyl groups of sialic acid at the suface. Besides the electrophoretic mobility, neuraminidase also alters the immunological reactivity of the cell since the M and N blood group antigens and heterophile antigens are inactivated [164, 633].

*Adenoviruses.* The mechanism of the hemagglutination reaction of adenoviruses has been little studied in biochemical terms. It was suggested on the basis of reactivity to blocking groups that nonterminal arginine residues on the fiber and glutamic or aspartic carboxylic acid groups on the cell may be required for the reversible interaction of adenovirus type 7 and erythrocyte membranes [477] and hopefully this may serve as a model for adenovirus-host cell interaction (sect. III).

*Picornaviruses.* There are two stages of interaction of enteroviruses with erythrocyte receptors. In the first, which can occur at 4 °C with some viruses, the attached virions can be recovered again in infective form. The second step leads to a loss of virus infectivity and it has been suggested that there is a parallel between these events in erythrocytes and in susceptible host cells [528]. In agreement with this, there is evidence that soluble purified erythrocyte receptors cause the release of infectious RNA from echoviruses [532]. This indicates that viral uncoating is membrane-mediated, but there may be an alternative less sophisticated explanation since highly purified echoviruses are fragile and can also spontaneously release nucleic acid [532]. Others have also claimed that they cannot observe the irreversible alteration of enteroviruses after interaction with erythrocytes [EGGERS, personal commun.].

# V. Irreversible Alteration and the Penetration of the Virion into the Host Cell

## *A. Introduction and Methodology*

Irreversible alteration of infective virions and the penetration of virions into the cell will be discussed together although they differ conceptually and may in some cases occur separately. In other cases, such as the proposed fusion of enveloped virions with the cell membrane, the two steps may be conceptually identical.

Virus which cannot be recovered in infective form after interaction with cells has been described as having undergone 'irreversible attachment' [523], 'eclipse' [200, 271], or preferably, 'irreversible alteration'. This reduction of recoverable infectivity signifies the entry of cell-associated virions into the 'eclipse' phase of the infection [407]. The biochemical nature of the alterations in the virions which lead to reduction in infectivity will be discussed in section VI. Some input virus also remains bound unproductively to cellular receptors, or is encapsidated in vesicles but is otherwise unaltered, and this usually accounts for the fraction of the parental infectivity which can be recovered at any state of the infection [229, 420 a]. In addition, the nucleic acid or subviral particles of many viruses have detectable infectivity and in one case, reovirus, the genome remains in a structure which retains demonstrable infectivity.

The method used to recover infective virus from cells determines the yield and hence the apparent rate of irreversible loss of infectivity. Acid-resistant picornaviruses, which have not penetrated the cell membrane, have been removed by treatment at pH 1.5–2.5 [200, 419, 443, 750] which also kills but does not lyse the cells. This reversibility resembles the acid reversibility of the virion-antibody bond [418], although the latter is stronger [271]. Other procedures used to release bound picornaviruses also lyse the infected cells; these include freezing and thawing [148, 150, 419], 6 M LiCl [271], 8 M urea [271], sonic disruption [148, 276], 0.4% sodium dodecylsulfate (SDS) [82, 83, 399, 419, 420, 677], or 2 mg/ml deoxycholate (DOC) [200]. Treatment of infected cells with SDS releases poliovirus which cannot be recovered with acid treatment [419]; treatment with acid releases poliovirus which has passed beyond the stage of recovery with DOC [200], LiCl or urea [750]. Freezing and thawing or sonic treatment are less effective. Trypsin treatment can destroy polio receptors before attachment with virus but it does not dissociate the virus from the complex with receptor once formed [271]; on the other hand, treatment with chymotrypsin can release at least some receptor-bound poliovirus [528, 748]. Detergents also effectively recover receptor-bound adenovirus [383, 401], SV40 [25] and reovirus [91].

Recovery of infectivity from enveloped virus-receptor complexes is difficult since agents which disrupt the cell membrane also usually disrupt the virion envelope. Cell-associated influenza virus can be dissociated by repeated washing [4] but is partly resistant to digestion with neuraminidase which destroys receptors in the absence of virus [4, 311]; NDV has been dissociated from cellular receptors with excess neuraminidase [177]. There are probably both neuraminidase-sensitive and neuraminidase-insensitive stages in the attachment of these viruses [293]. Vaccinia can be recovered from cells by sonication [6]. Herpes simplex can attach to and be removed by washing from methanol-fixed cells but it cannot be removed from viable cells by washing, even after attachment at 4 °C [297]. Specific dissociating reagents have been described such as heparin sulfate for herpesvirus [267] and streptomycin, protamine or high salt concentration for arboviruses [461].

'Penetration' of virus into the cell, another step of early interaction, is also defined in operational terms. It is evident that the virus, or at least the virus genome, must penetrate into the host cell but the methods for detecting penetration are equivocal. Penetration has been measured by the loss of antibody-sensitivity of infective centers [150, 269, 276, 297, 311], but it is possible that the antibody-sensitive step in the infective process can occur outside the cell. For example, native picornavirus virions adsorb neutralizing antibodies while irreversibly altered 'A particles', produced from these by the eclipse reaction (sect. VI), do not [402]. A particles located on the exterior surface of the cell would thus not be sensitive to antibody neutralization and if they were in the infective pathway they would appear to have 'penetrated' by the criterion of loss of antibody sensitivity. Furthermore, it is possible that virus attached to cells may be shielded from antibodies by components of the cell's surface. For example, fully infective phage ØX174 cannot be neutralized by antibody after it has attached to the *E. coli* cell wall [479] and in such a case the kinetics of the development of antibody resistance would also provide a false measure of penetration. In other cases, animal viruses reacted with neutralizing antibody may show a transient enhancement of infectivity which is only lost upon prolonged incubation [257]. On the other hand, it is also remotely possible that antibody may be able to penetrate the cell membrane in some cases although attempts to detect this have failed [506]. Similarly, the inactivation of virus-cell complex at the cell surface by chemical means, such as acidification [297, 343], cannot be used if irreversible alteration of the virus at the cell surface is a part of the infective path, unless extracellular 'eclipsed' virions are further damaged by this treatment. The agent used also must not penetrate the cell. The electron microscope is another approach with limitations, since virus particles can appear to be within the cell because of artifacts arising from the location of the cell surface in respect to the plane of the section; other artifacts have also been discussed [139, 451].

### *B. The Effect of the Cellular Environment*

Irreversible alteration of infectivity of receptor-bound virus is temperature-dependent. Both eclipse and penetration are inhibited at low temperature for poliovirus [271, 419], EMC [244] and adenovirus [383]. This effect is also seen with the irreversible alteration of picornavirus by soluble receptors *in vitro* [528]. In the case of HRV2, a 6-fold decrease in the rate of eclipse was observed when the temperature was decreased from 34 to 17 °C, a factor which is similar to the decrease in rate of attachment in this system [Lonberg-Holm, unpublished]. It is, however, generally possible to attach virus to cells at low temperature in order to block subsequent steps leading to irreversible loss of infectivity. Where the rate of attachment is temperature-sensitive, however, attachment should be carried out at a high cell density to overcome the decreased rate at low temperatures. With this technique, the eclipse of poliovirus by HeLa cells was shown to have first order kinetics at 37 °C with a half-life of 20 min; penetration to antibody resistance also showed similar kinetics [271]. Poliovirus penetration (resistance to neutralizing antibody), eclipse (loss of SDS-recoverable infectivity) and uncoating (development of photoresistance of virus containing neutral red) has been studied carefully as a function of temperature. Mandel [423] found that the rate of penetration proceeds faster than the irreversible loss of infectivity or uncoating at 22–25 °C.

The effect of pH on the irreversible alteration of enterovirus has been examined in a few isolated instances. The optimum pH for attachment of coxsackievirus B3 to HeLa cells is 4.5 while the optimum pH for either irreversible alteration or elution is about 7 [126]; the rate of eclipse of echovirus 7 by intact erythrocytes and soluble erythrocyte receptors is also maximal at pH 7.6–8.0 [528, 532]. There is indirect evidence that $Mg^{++}$ is required for penetration and irreversible alteration of poliovirus by monkey kidney cells while these steps were enhanced by lowered $K^+$ [357].

Penetration and/or irreversible alteration of influenza virus [311], vaccinia virus [6] and herpes simplex virus [194, 280, 297] is temperature-dependent and presumably all enveloped viruses show such a dependence. The effect of ionic strength and medium composition has not been examined systematically but penetration and/or eclipse of Sindbis virus is enhanced by reduced ionic strength [453].

### *C. Inhibition or Stimulation*

Different processes at the cell surface are probably involved for different viruses; some evidence for this specificity is provided by differential sensitivity to inhibitors.

Chemical inhibitors of irreversible alteration and/or penetration can be arbitrarily divided into two classes, those which interact with virions so as to *stabilize* them against processing and those which react with a cellular component involved in the early steps of infection. The first type of inhibition is illustrated by the prevention of elution of altered and noninfective poliovirus from host cells by 10 mM glutathione [200]. This elution is now known to be a concomitant of the irreversible alteration of poliovirus (see below). A significant retardation of eclipse of poliovirus by 1 mM glutathione has also been demonstrated [354]. This may be related to an enhanced stability of poliovirus against heat inactivation in the presence of $10^{-2}$ M glutathione [LONBERG-HOLM, unpublished].

The antiviral activity of certain compounds appears to be related to their ability to stabilize virions against *in vitro* and *in vivo* alteration [669]. Rhodanine blocks infectivity and uncoating of echovirus 12 and also stabilizes it against alkaline disruption *in vitro* [181]. SDS at 30 μg/ml specifically blocks human rhinovirus type 2 infectivity and cell-mediated irreversible alteration and also stabilizes virions against acid and thermal inactivation [400]. SDS also stabilizes the native surface conformation of a subpopulation of HRV2 natural top component particles which share antigenic determinants with native virions and which attach to virus-specific receptors (sect. VI). Thus SDS appears to interact with the virion proteins to stabilize their native conformation. Sodium salicylates show antiviral activity against EMC virus which may be caused by their ability to slow irreversible alteration and/or penetration [309]. It is unclear whether the salicylates affect the virions or the host cells.

'Symmetrel'® (aminoadamantane) appears to block infectivity of influenza virus by preventing penetration of virus into the host cell, as determined by its effect on the development of antibody resistance during the formation of infective centers and loss of hemadsorption due to infecting virions [154, 269]. Aminoadamantane may also prevent uncoating of influenza (fowl plague) virus, rather than penetration, based on the photosensitivity of neutral-red-labeled virus [337]. It may not be possible with enveloped viruses, however, to separate this type of uncoating and irreversible alteration, since both may be initiated by an opening in the virion envelope. Aminoadamantane or related compounds also have antiviral activity against Rous sarcoma virus [709], rubella virus [509], Sendai and pseudorabies viruses [474], but not herpes simplex [509] or various non-enveloped viruses.

Certain toxic sulfhydryl reagents including $HgCl_2$ can inactivate vaccinia and fowl plague virus and the effect is reversible by glutathione [6]. Such inactivated virus appears to attach to and penetrate into host cells but is not uncoated [6]. In this the sulfhydryl reagents may have a mode of action which is formally analogous to that of aminoadamantane. Certain carbobenzoxytripeptides are inhibitors of measles virus and they have also been shown to block virus penetration and virus hemolysis but not virus attachment [494]. Antiviral compounds which block irreversible alteration by direct interaction with virions often show narrow spectra of activity and this is probably because virions differ from each other in the details of the structure of their envelope or of their capsid surface (sect. II. A).

The second site for inhibition of irreversible alteration and penetration of viruses is the cellular structure involved. Holland [271] found that the irreversible alteration of poliovirus was remarkably insensitive to metabolic inhibitors such as azide, fluoride or DNP, as well as inhibitors of protein synthesis. Accordingly, eclipse and/or uncoating of enteroviruses have been observed *in vitro* with erythrocyte membrane fragments (sect. IV). This lack of an energy requirement may however be fortuitous since *in vitro* systems may not parallel *in vivo* systems and since stringent inhibition of the host's metabolic energy production might inhibit eclipse as shown for the bacteriophage R17 [146]. Irreversible alteration of influenza virus readily occurs *in vitro* with membrane fragments but here there is a dependency on added ATP (see below).

Con A attaches to cell membranes (sect. II. B) and membranes of enveloped viruses [30, 501, 507]. It was recently shown that Con A can inhibit the infective process of Sendai, herpes and, to some extent, poliovirus [501]. In the first two cases, Con A interacts directly with the envelopes of the virions and in all 3 cases, Con A interacts with the membrane of the host cell. The enveloped viruses may thus attach to the cell through Con A linkages, and then be unable to proceed with the infective process. Poliovirus, however, does not interact with Con A; in this case and also in the case of human rhinovirus type 2, both attachment and the cell-mediated irreversible alteration of infectivity are inhibited by the lectin, presumably because Con A interacts with the cell membrane [LONBERG-HOLM, unpublished]. Although there are fewer sites available for adsorption of rhinoviruses in the presence of Con A, a major effect also appears to be inhibition of the reaction leading to noninfective subviral particles lacking VP4. The latter inhibition may result from the ability of multivalent Con A to cross-link and block diffusion of membrane components (sect. II. B), thereby keeping the virus receptor complex isolated from components needed for irreversible alteration. Alternatively, since Con A inhibits phagocytosis [37], penetration of virions by pinocytosis may also be blocked.

It should also be noted that some substances may enhance irreversible alteration or penetration of viruses and hence enhance infectivity through an effect on the cell membrane. The alkaloid cytochalasin D enhances infectivity of poliovirus, but not vaccinia virus, at an early stage subsequent to attachment [157], and this effect is prevented by dibutyryl cyclic-AMP [158]. A closely related alkaloid, cytochalasin B, inhibits both sugar transport [190] and cell membrane mobility [79]. The cytochalasins may [8, 153, 703] or may not [732] inhibit pinocytosis, and they may eventually be useful in the delineation of the relative importance of the different pathways for virus penetration.

Cardiac glycosides have been noted to enhance the size of EMC plaques [717]. Cardiac glycosides also have been shown to enhance poliovirus infection of monkey cells and evidence has been presented that this effect is on the penetration step or a step directly after attachment [238, 355, 356, 358, 360].

### *D. The Effects of Antibodies*

Antibodies inhibit the post-attachment steps of infection in one or two ways. First, antibodies may react with the virions, either before or after attachment to host cells. Virus which has been 'neutralized' by antibody can often attach normally to host cells as shown for poliovirus (sect. III) [417, 419, 421]. When poliovirus is neutralized, it is unable to proceed beyond the acid-elutable stage of interaction [419]. It should be noted that acid pH dissociates antibody-virus complexes as well as cell-virus complexes [417–419, 643]. Neutralized poliovirus, furthermore, does not undergo irreversible alteration leading to a subviral product which resembles the normal product whose RNA can be uncoated by trypsin (sect. VI) [275]. Antibody therefore appears to 'lock' the virion capsid into a conformation which prevents the normal sequence in infection [395a] and also prevents a pH-dependent reversible alteration of intact virions, as detected by isoelectric focusing [424]. Neutralized poliovirus [422] and adenovirus particles are eventually [526] digested by the host cell and some of the breakdown products are eluted back into the medium. Neutralized vaccinia virus is also preferentially sequestered and destroyed in macrophage phagosomes [617] and it is tempting to attribute these effects to an enhancement of a lysosome-mediated host defense against virus invasion.

The bond between antibody and a nonenveloped virus is not readily reversible at neutral pH [418, 524], but virus infectivity can be recovered by treatment with acid (above) or by other means [425; review of literature, 658], and it should be possible to make a wider examination of the extent to which neutralization results in the stabilization of the virus capsid conformation. It should be noted that monovalent antibody fragments, although able to produce neutralization, are far less efficient than bivalent fragments both for poliovirus [338] and for phage f2 [168]. Also, although many antibody molecules may bind to any given virion, the kinetics of neutralization are pseudo-first order with antibody concentration which implies that one single antibody molecule may be capable of producing neutralization [168, 417]. However, in practice, many antibodies attach to a virion before it is neutralized [524]. If the effect of antibody is to 'lock' the virion into a conformation which cannot undergo essential steps in the infective pathway, it is reasonable that the host cell should influence the extent of neutralization. Different host cells may also differ in sensitivity for the inhibition of this reaction [395a].

Early studies with influenza virus suggested that neutralized virions could be blocked in attachment, or in ability to penetrate into the host cell [303]. The kinetics of neutralization of enveloped viruses indicate that neutralization is reversible [197, 198] or partly reversible after short periods of interaction [11, 374]. The interaction of antibody with the cell membrane has also been reported to be reversible, with an average Ka of $10^7$ [51]. Complement may constitute a third component in the process of neutralization of enveloped viruses [11, 551, 710]. Equine arteritis virus requires complement for neutralization and sensitized virus which has not reacted with complement attaches to and penetrates host cells in a normal fashion [551].

A second site for antibody to inhibit the irreversible alteration or penetration of virus is by direct reaction with the surface of the host cell. The ability of anticellular antibody to inhibit attachment of virus has already been discussed in section III; other steps in early interaction may also be inhibited. Anticellular antibodies inhibit echovirus 1, but attachment of virions and the infectivity of virus RNA were both unimpaired [682]. Thus penetration or irreversible alteration or uncoating might have been inhibited and the inhibitory effect might be analogous to the effect on rhinovirus type 2 of Con A (above) which, like antibody, is a large polyfunctional protein. The effect might be mediated by an inhibition of either the activity or mobility of a component in the cell membrane, or alternatively by inhibition of pinocytosis [78a]. Anticellular serum has also been reported to inhibit penetration of vaccinia virus [617]. Since the reaction of antibody with cell membrane components is probably selective, different viruses, which probably employ different cell membrane components for their early steps in infection, will be inhibited to a varying extent.

### *E. Elution versus Penetration*

Elution of virus can take two forms. In the first of these, infectious virus is released from host cells after it has attached; this has been discussed in section III. A second and more interesting phenomenon is the elution of noninfective virions after interaction with the cell; this may be viewed as an alternative to penetration. Elution of altered virus is a significant feature of the early interaction of many picornaviruses and has been examined intensively for poliovirus type 1 [200, 321, 422], coxsackievirus B3 [126, 127] and rhinovirus type 2 [399, 488]. The eluted particles sediment at about 90 % the rate of infectious virions, have altered surface structure but contain RNA; the biochemical nature of the eluted particles is probably the same as that of the cell-associated, irreversibly altered virus. The eluted poliovirus [200] and coxsackievirus [126] particles also appear to have bound a fragment of the cell membrane. There is also a relationship between cell-associated eclipsed bacteriophage ØX174 particles and the noninfective eluted ØX174 particles [479], and some cell-associated phage R17 also elutes as noninfective particles [513].

It has been assumed that the elution of picornaviruses is a concomitant of the reaction leading to the eclipse of virus by the cell [126]. Both elution and cell-mediated irreversible alteration of poliovirus are first-order reactions [321]. Also, elution and eclipse of coxsackie B3 are both pH-dependent in a similar manner [126]. Elution of noninfective poliovirus is unaffected by changes in pH between 5.5 and 7.9 [200] and by addition of EDTA [200, 321] but the extent of elution probably depends upon the nature and the physical state of the cells as well as the virus used. Up to 80 % of cell-associated labeled poliovirus [321] or coxsackie B3 virus [127] can elute after incubation at physiological temperatures while type 2 rhinovirus usually elutes only to about 20 % [488].

There are at least two ways to view the elution of noninfectious virions. First, if all virions are altered irreversibly outside the cell and only then may penetrate through the membrane directly, some also may detach, depending upon random choice or the location of the particle (fig. 6, reaction 3a). Secondly, if virions are engulfed before irreversible alteration (fig. 6, reaction 3b), those particles within vesicles will be unable to elute while those outside may do so.

The elution of 14S capsid fragments as found with EMC and other cardioviruses [244], and 12S fragments from FMD virus [67, 667], is analogous to elution of relatively intact virions. In the case of the cardioviruses, elevated pH or $MgCl_2$ concentration increases the efficiency of infection and acts to prevent elution of 14S protein units [244]. It has also been found that 5–8 % of the capsomeres of adenovirus type 2 elute during the course of infection [525].

### *F. Penetration versus Pinocytosis of Nonenveloped Virions*

There has been a continuing difference of opinion concerning the mechanism by which animal viruses or their genomes pass into the host cell [138, 139, 523]. One aspect of this problem is semantic, since a virion which is within a vesicle is still outside the cell's cytoplasm and in a topological sense is also outside the cell. Inert particles taken into pinocytotic vesicles do not enter the cytoplasm [186] and it is evident that the virion must employ a special mechanism to pass the barrier of the cell membrane.

Figure 6 shows several models for the penetration of nonenveloped viruses. The first line represents the simple penetration of the genome as uncoated nucleic acid. A similar process probably occurs in bacteriophage infection although highly specialized structures are involved (e. g. the T-even phages). Before the nucleic acid of spherical phages penetrates, the attached particles may undergo a conformational alteration [308, 479, 513] which resembles the irreversible alteration observed with animal viruses; the elution of noninfective phage may also be a concomitant of this alteration. The irreversibly altered particles in figure 6 are shown with a wavy outline to indicate altered surface structure. The process depicted in figure 6, line 1, probably does not occur as such with animal viruses, although there is a possibility that the genome of the cardiovirus may gain entry by a similar process [244].

Another possibility, ably defended for 15 years, is that a virus must gain entry into the animal cell by pinocytosis, 'colloidopexis' or 'viropexis' [139, 195]. This process is depicted, in part, in figure 6, line 2, and figure 7, line 1. In figure 6, line 2, we have indicated that virus can undergo irreversible alteration, shown as a change in surface structure, either before (2a) or after release from the pinocytotic vesicle (2b). We have indicated release to occur by dissolution of the vesicles.

Perhaps the best established case for the role of pinocytosis in the infective process has been made for reovirus on the basis of ultrastructural and biochemical evidence. Reovirus attaches to L cells, enters 'phagocytic' vesicles and is eventually sequestered in lysosomal vesicles and the virion protein is then partly digested [618]. Reovirus may, however, be unique in requiring proteolytic digestion because its double-stranded RNA genome remains undegraded after prolonged sojourn in the lysosome. Even in the case of reoviruses, however, it has also been suggested that virions can enter the cell by direct penetration as well as by pinocytosis [455].

It has also been suggested that adenoviruses may enter via pinocytosis [92, 93, 136, 203]. In an early electron microscopic examination of the entry of adenovirus type 7 into HeLa cells, Dales [136] found virus particles densely packed in lysosomes and gaining relatively slow entry into the

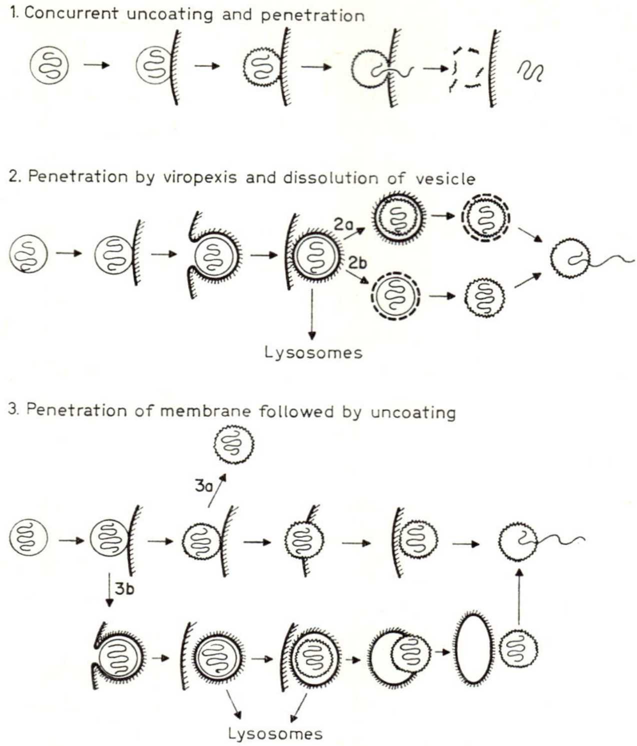


*Fig. 6.* Schematic representation of some alternative pathways for irreversible alteration, penetration and uncoating of nonenveloped viruses. Surface alteration of virions is indicated by a change from smooth to wavy outline of the capsid. See text for further details.

cytoplasm. Biochemical studies, however, later indicated a rapid entry of types 2 [401, 525] and 5 [383, 656] into the cytoplasm; a subsequent electron microscopic analysis of type 5 [92] and also types 1, 7 and 12 [93] confirmed the rapid entry of at least some adenovirus virions into the cytoplasm, with significant quantitative differences in the behavior of the different serotypes. Quantitative electron microscopic investigation revealed an apparent sequence leading from attachment to intravesicular particles and then to cytoplasmic particles; this process required about 1 h for adenovirus type 5 [92]. Some particles also passed into lysosomal vesicles, and this was more evident for adenovirus type 7 than type 5 [93]. Chardonnet and Dales [92, 93, 94] concluded that virus gained access to the cytoplasm by pinocytosis.

Another possibility for the entry of nonenveloped viruses is illustrated in figure 6, line 3. In this model, virus attaches to the cell surface and is altered by the cell to permit penetration of the membrane, as indicated by the altered surface of the particle. The reaction shown as 3a indicates the elution of noninfective altered virions such as are found with poliovirus. A possible modification of the direct penetration sequence is indicated by reaction 3b. In this case, virus is taken into pinocytotic vesicles and there undergoes alteration permitting penetration of the membrane of the vesicle. Such a pathway might prevent excessive 'elution' of altered virus.

Some evidence has been found that adenoviruses may gain entry by penetration of the cell membrane in the absence of pinocytosis. Adenovirus type 2 virions are rapidly altered to a partially DNase-sensitive particle (sect. VI) so that the half time for intact cell-associated virions is only about 15 min [401]. Intracytoplasmic particles of adenovirus type 7 [459] or type 5 [92] are also seen by electron microscopy, within 10 min after interaction with cells at 37 °C. These intracellular particles may also have been structurally altered [459]. Intracytoplasmic adenovirus particles have a transient life in the cytoplasm and become further altered at other locations, mainly in or at the nuclei [92, 93, 401, 459]. Ultrastructural studies of this process for adenovirus type 7 led MORGAN *et al.* [459] to propose that entry occurs in the absence of pinocytosis. Very recently BROWN and BURLINGHAM found further evidence that the primary route of penetration of adenovirus type 2 is directly through the plasma membrane; virions traversing the plasma membrane were detected in replicas of freeze-etched infected cells [66a]. In the ultrastructural studies showing direct penetration of adenoviruses, virus particles were also detected in intracellular vesicles, but with a relatively low frequency. In addition to the evidence that pinocytosis may be unnecessary for penetration it is also generally accepted that entry of adenovirus into a lysosomal vesicle results in the ultimate destruction of the virion and the abortion of the infective process.

Evidence of an indirect nature may support the contention that poliovirus enters the cell by pinocytosis. Penetration and irreversible alteration of poliovirus occur both at the same rate at 37 °C [271, 423], but at 25 °C infective centers escape neutralization before all cell-associated infective virus has been lost [423]. The loss of sensitivity to neutralization before eclipse might result from the tight binding of virus to the cell membrane, but is more likely attributable to the entry of virions into pinocytotic vesicles (fig. 6, line 2 or reaction 3b). However, it must be borne in mind that infective centers might escape from neutralization before penetration, as discussed above, and also that not all cell-associated infective virus need be located on a pathway leading to an infective center. For example, some infective virus might be sequestered in lysosomal vesicles awaiting ultimate digestion and yet be recovered after cell lysis.

Electron micrographs of cells infected 2, 5, or 10 min with poliovirus appear to show direct penetration through the cell membrane [176]. Susceptible cells of 3 types were examined and virus-like particles were found free inside the cytoplasm early in infection and the intracytoplasmic particles may have undergone a morphological alteration. At the present time, there is no solid evidence which rules out any of the processes shown in lines 2 and 3 of figure 6.

Further insight into the penetration of picornavirus may be obtained with inhibitors. Pinocytosis may be inhibited by poisons of respiration or glycolysis [105] but the penetration of poliovirus does not appear to be sensitive [276]. The effect of cytochalasin B and D on the fate of poliovirus in early infection needs further investigation (see above). The effect of divalent cations also should be studied since $Ca^{++}$ has been reported to be essential for phagocytosis [550a].

An unusual process has been described for the penetration of simian virus 40, a papovavirus, into host AGMK cells [305]. Electron micrographs indicate that the virions are tightly enveloped by the cell membrane (these are represented as X in table IV, sect. VI). These 'enveloped' virions then migrate to, and fuse with, the nuclear membrane to release free virions again within the nucleus. It is not known where and when irreversible alteration of the virions occurs; *uncoating* occurs after penetration into the nucleus [25]. Polyoma virus may also penetrate into the cell with an envelope composed of the plasma membrane [438].

### *G. Penetration versus Pinocytosis of Enveloped Virions*

The mode of entry of enveloped viruses into the host cell will be discussed in relationship to figure 7. The nucleoprotein capsid represented as an oval may constitute either a spherical capsid, as with herpesvirus, or a helical nucleocapsid as with myxovirus. The major alternatives are viropexis, line 1, and the fusion of virion and cell envelopes, line 2. In line 1, viropexis leads first to lysis of either the virion envelope or the vesicle. In line 2, the various possibilities for the steps leading to uncoupling and recoupling of the virion and cell envelopes are not shown. One model has

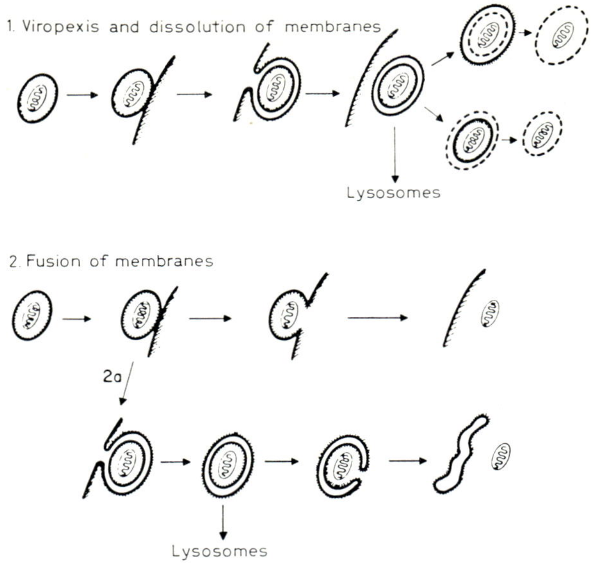


*Fig. 7.* Schematic representation of some alternative pathways for penetration and partial uncoating of enveloped viruses. A similar schematic was shown by DURAND *et al.* [177]. See text for further details.

been proposed for cell-cell fusion induced by lysolecithin [402a]. The evidence either for pinocytosis or for fusion is voluminous and it is impossible to decide between the alternatives simply on the basis of the number of references for each view. We will discuss below the status of our understanding of penetration for some enveloped viruses.

The virions of poxvirus should be ideal subjects for investigation because they are large and morphologically distinct and because they show a ratio of PFU to physical particles approaching unity. Vaccinia infection shuts down host protein synthesis within 20 min after infection [47, 189], indicating that penetration must be very rapid. Authoritative work [137, 141] indicates that vaccinia enters the cell rapidly by endocytosis and that virion cores are found in the cytoplasm 20 min post-infection. The uncoating of intravesicular virus and the egress of cores from the vesicles must be rapid because the steady-state number of intravesicular virus particles was low but the mechanism for uncoating and penetration into the cytoplasm was not resolved. Fluoride, which inhibits phagocytosis, also causes most of the cell-associated virions to remain at the cell surface in an antibody-neutralizable location [141]. Heat-inactivated or antibody-neutralized vaccinia can be taken up by pinocytosis but is not released from vesicles so that cores are not found in the cytoplasm and such virus is eventually digested. Aggregated infective virions are also engulfed by the cell *en masse* but are subsequently uncoated in a normal fashion [141].

On the other hand, a poxvirus of insects penetrates the insect intestinal epithelium by fusion with the membrane of the microvilli [233]. In this case the microvilli have an even smaller diameter than the poxvirus cores. Fusion of vaccinia virus with the membrane of mouse L cells has also been observed, but only after 4 h of infection, and fusion of this type is said to represent an aberrant stage associated with a few particles remaining immobilized at the cell surface [139]. The baculoviruses (insect nuclear polyhedrosis virus and granulosis virus) contain large bacilliform nucleocapsids within envelopes which also fuse with host microvilli, as shown by independent work [251, 653, 654]. Other work with this family of viruses indicates that viropexis may also occur when virus attaches to less specialized cell surfaces [670].

The penetration of herpesvirus has been interpreted in alternate ways. Earlier electron microscopic studies concurred that herpes virions enter the cell by pinocytosis [186, 280, 615]. Colloidal gold was employed for the identification of pinocytotic vesicles and virion capsids were found to pass out of the vesicles while gold particles of a similar size did not [186]. Capsids without envelopes were detected to a small extent inside and to a large extent outside the vesicles but the mechanism of egress was not detected. Dales and Silverberg [142] also confirmed the rapid rate of pinocytosis of herpesvirus but detected relatively few free intracytoplasmic virion capsids. Intracellular nucleocapsids may also be derived from disrupted virions in the inoculum because the nucleocapsids themselves are also infective to a small degree [2, 631].

Morgan *et al.* [458] also found herpes virions in phagocytic vesicles during early infection but suggested an alternative mode of entry. They contended that the virion and cell envelopes fused as a result of enzymatic alteration of both membranes at the site of attachment and were unable to detect passage of virion capsids from vesicles into the cytoplasm, even by the process depicted in figure 7 beginning at reaction step 2a. Others may, however, have visualized such a process [306]. Recent work emphasizes the speed of fusion of virus and cell surfaces; nucleocapsids are found in maximal concentration in the cytoplasm after 10–20 min and are subsequently converted to empty capsids [452]. Infectious bovine rhinotracheitis virus, another member of the herpes group, also undergoes fusion with host cell membranes [753]. Morgan *et al.* [458] concluded in 1968 that 'electron microscopic examination alone cannot define which of two viral particles entering by two different mechanisms actually infects the host cell' and that other methods are needed to resolve the problem. It may be significant that cell fusion and polykaryon formation is a feature of herpes infection, and the process leading to the fusion of infected cells may also be related to the process leading to fusion of virus and cells. This parallel has been reviewed [486].

The penetration of leukoviruses deserves further investigation. It has been suggested on the basis of electron microscopy and radioautography that the avian viruses RAV-2 and SR-RSV enter by viropexis and subsequently migrate to the nucleus where virus-specific RNA and DNA synthesis occurs [140]. Earlier work with murine leukoviruses suggested that reverse transcriptase activity is in the cytoplasm [256] and also that there is direct penetration of intact enveloped and de-enveloped virions [451]. This direct entry occurred in a way which differs from the mechanisms shown in figure 7. The cell membrane was altered or digested at the point of virus attachment and the intact particle invaded the cytoplasm [451]. Dissolution of both cell and virion membrane also occurred, in which case the virus nucleocapsid alone penetrated the cell. No fusion of the virion and cell membranes was detected which correlates with the lack of polykaryocyte formation in leukosis-infected cells. Penetration was maximal 30–45 min after infection, whereas some phagocytosis was detected after only 5 min.

The fusion of NDV or Sendai virions with the membranes of host cells has been demonstrated unambiguously [161, 449, 456]. This occurs rapidly, within 2 min after the start of incubation [393, 456]. Evidence for the fusion of virion and erythrocyte membranes has also been presented [17, 17a, 292] and virion envelope antigen can diffuse outward over the cell surface from the point of fusion [17]. Radioautographic studies also indicated that virion protein remained at the cell membrane following penetration of virion RNA [755]. On the other hand, it has also been reported that NDV enters host cells by viropexis [288, 619], although results from one of these studies also shows fusion of the membranes of adjoining cells in the region of attached virus [288]. Evidence of an indirect nature has also been found for the participation of pinocytosis in the infective process of NDV [177].

Paramyxovirus virions can produce either lysis or cell fusion and these abilities are probably related [292]. The fusion of cell membranes to the membranes of virions or other cells during polykaryocyte formation may be viewed as a three-step process: the attachment of adjoining envelopes, formation of lesions by a 'disconnecting agent', and then the repair of these lesions. The 'disconnecting agent' may either differ from neuraminidase [754] or be related to neuraminidase [292] and metabolic activity may be required for repair in some cases [503] but not in others [292]. Cell fusion requires $Ca^{++}$, and removal of $Ca^{++}$ results in lysis without fusion [502]. Although polykaryocyte formation can be induced in uninfected cells by lysolecithin [542], analysis of virus-treated cells or of parainfluenza 2 virions themselves has failed to detect the presence of lysolecithin or of phospholipase [183]. However, it was reported that enzymatic removal of lysolecithin reduces syncytia formation by Sendai virus [24].

The penetration of myxoviruses has been attributed to pinocytosis, and the term 'viropexis' was originally introduced by FAZEKAS DE ST. GROTH [195] to describe pinocytosis of influenza virus by the cells of the chorioallantoic membrane. Virion antigen, detected by its ability to adsorb antiviral antibody, was lost after introduction of viruses into the egg and this was attributed to ingestion of the entire virion. Recent ultrastructural studies also support the pinocytotic entry of influenza into ascites tumor cells [16]. Mouse hepatitis virus (a coronavirus) has also been reported to enter the host cell by a process of pinocytosis [152].

On the other hand, evidence has also been accumulated in favor of the entry of influenza by fusion. In 1962, HOYLE et al. [294] found that non-nucleated fragments of chick chorioallantoic membrane cells could interact with and labilize virions so that they lose their inner nucleoprotein component. The *in vitro* interaction of plasma membranes from chick embryo cells with influenza virus leads to a depression at the site of attachment and then, in the presence of added ATP, a disintegration of both virus and cell membranes and release of virus nucleoprotein [100, 365]. Influenza virus was also found to fuse with cells of the chorioallantoic membrane after about 5 min of incubation; either the virion membrane or the cell membrane appeared to open first but in either event, the lesions were rapidly repaired since 'cells which must have received many viral particles showed no visible alteration after the process of entry had been completed' [457]. The same process has also been seen in infection of chick embryo tracheal cells [44].

No evidence for fusion of VSV virions and mouse L cells was detected in a careful study in which a large number of sections were examined 0, 1, 3, 5, 10 min or more after cells infected at 4 °C were warmed to 37 °C

[622]. However, 50% of the virus entered pinocytotic vesicles within 5 min; the mechanism of intracellular release from pinocytotic vesicles was not determined. Rabies virus has also been reported to penetrate by viropexis [304]. An independent investigation of VSV penetration unfortunately employed high speed centrifugation for the attachment of virus to cells which prohibits direct comparison [261]. The results indicated that virions fused with plasma membrane so that virion envelope proteins remained outside and nucleoprotein entered the cytoplasm [262].

### *H. Irreversible Alteration or Penetration as a Locus for Host Resistance*

In some nonpermissive systems, virus can attach normally to the cell but a subsequent reaction leading to irreversible alteration or penetration is defective and this defect is the basis for the host's resistance. Ten years ago, HOLLAND [271] found that a line of HeLa cells could adsorb but not efficiently eclipse poliovirus types 1 or 2 or coxsackievirus B1. Other HeLa cell lines had been found at an earlier date to contain poliovirus receptors yet to be comparatively resistant to poliovirus infection [150, 695]. In one case, both penetration (as based upon establishment of antibody resistance) and the susceptibility of cells to infection with RNA appeared normal [150]; we do not know if irreversible alteration was blocked. SV40-transformed human amnion cells were also found to be resistant to polioviruses but could be infected with poliovirus RNA [241]. Virions could both attach and penetrate to an antibody-resistant state at a normal rate and it was concluded that an uncoating step was defective. The data, however, also indicate that irreversible alteration was at least partly blocked. Small leukocytes also appear to be relatively resistant to poliovirus until they are stimulated to undergo blastogenesis by phytohemagglutinin [729, 731]; the block in infection may be both at attachment and at irreversible alteration [729]. It has also been reported that SV40-resistant monkey cells are defective in the penetration or uncoating of SV40 [558] since virus attaches normally but virion DNA does not enter the nucleus. The SV40-resistant cells are subdiploid but their chromosome number increases with passage as does their susceptibility to SV40 [559]. An enhanced ability of fused hamster cells [472] or of hamster and chick heterokaryons [184] to support polio replication have also been viewed as resulting from enhanced penetration during fusion. A complicated system has also been described in which the efficiency of infection of a resistant line of HeLa cells with EMC is enhanced by concurrent infection with poliovirus [612]. The helper poliovirus does not increase the efficiency of either attachment or irreversible alteration and it was suggested that uncoating was enhanced.

In other nonpermissive systems for nonenveloped viruses, cellular resistance is not attributable to restricted attachment, eclipse, penetration or uncoating [71a, 136, 203, 204, 649, 706, 707].

With enveloped viruses, it has been reported that leukoviruses [534] and herpesviruses [673] do not penetrate certain resistant cells. In the latter case, feline herpesvirus could be induced to penetrate human cells by fusion with inactivated Sendai virus, but there was still no yield of progeny.

# VI. Uncoating of the Viral Genome

## *A. Introduction*

### 1. Criteria for Uncoating

It is now certain that parental viral nucleic acid need not be entirely uncoated of virion proteins to engage in transcription (i. e. vaccinia virus) or even in the events of replication (i. e. reovirus). Nonetheless, in many cases, the end result of the early steps in infection may be the uncoating of the genome so it can serve as a template in both translation and replication or transcription and replication. However, nucleic acids do not exist in an entirely free form within the cell and are probably always associated with cellular [295] or viral proteins. With the enveloped viruses, the process of removing the nucleoprotein from the envelope may also be described as 'uncoating' but the product is still an encapsidated nucleic acid.

Uncoating has been defined operationally in one of four ways. The first is by morphological alteration seen in the electron microscope. Thus, the dense 'nucleoid' of herpesvirus [458, 753] and the nucleoprotein core of adenovirus [459] have been visualized within the cell in the process of emerging from the virion capsid. Other criteria for uncoating are less direct: the second involves the use of photosensitive virus which has incorporated dyes such as acridine orange, proflavine or neutral red during growth. It has been assumed that the dye is released when the nucleic acid is separated from virion proteins. However, there may not be a direct relationship between the loss of photosensitivity and uncoating since either loss of the virion envelope or loosening of the capsid structure might also release the dye. A third approach involves the measurement of nuclease sensitivity of the nucleic acid of the infecting virions after the disruption of the host cell. This requires the use of isotopically labeled virions and high multiplicities of infection. The method has been used extensively with nonenveloped viruses, but the nature of the nuclease-sensitive products should be established. This can be overcome in the fourth approach, which entails the isolation and identification of uncoated subviral components.

## 2. Cellular Site of Uncoating

The plasma membrane, the lysosomes and the nuclear membrane often appear to be involved in uncoating but their roles differ from virus to virus. This has been reviewed extensively [139], and is summarized in table IV. In table IV, the letters A, B and C designate products derived from parental virions, increasing in extent of uncoating.

Picornaviruses are probably irreversibly altered at the plasma membrane and transformed to A particles which lack VP4 but still contain a full complement of RNA (see below). The fate of A particles within the cell remains to be established. Eluted A particles of poliovirus [200] and coxsackievirus [127] are bound to membrane fragments and there is indirect evidence that cell-associated rhinovirus A particles are also membrane bound [399]. Picornavirus RNA appears also to be replicated at a membrane site [76] but we do not know if it is carried there by A particles. The capsid protein of A particles is sensitive to proteolytic digestion [90, 275] and those A particles which find their way into lysosomal vesicles are probably uncoated; it seems likely, however, that the genome is also destroyed there by nucleases and digestion does appear to be the fate of most of the

*Table IV*. The cellular sites of uncoated products of different viruses

| Cellular compartment | Virus | | | | | | |
|---|---|---|---|---|---|---|---|
| | picorna | reo | adeno | herpes | papova | pox | myxo |
| Plasma membrane | A | | A? | | | | |
| Lysosomes | | A | | | | | |
| Cytoplasm and reticulum | A<br>B? | A? | A | A | X | A<br>B | A |
| Nuclear membrane | | | A<br>B | A?<br>B? | X | | |
| Nucleus | | | B<br>C | | virion<br>A | | (orthomyxovirus)<br>B? |

Modified from DALES [139]. See text for identity of components. In each case virus first attaches to the plasma membrane and component A is the first product of uncoating. Component X represents virions within a membrane vesicle.

picornavirus particles which attach to and remain with the host cell [321, 399]. On the other hand, reovirus appears to require a sojourn in the lysosome for uncoating as a part of its infective pathway [618] but its subsequent fate is not known.

Adenoviruses clearly undergo steps of uncoating at different cellular locations [92, 401, 459]. Virions first lose their vertex capsomeres at the plasma membrane (component A) and are then released into the cytoplasm. They then migrate to the nuclear membrane where the core (component B) is removed from the remaining capsid; this is illustrated in figure 8. The proteins of the core are removed later, presumably within the nucleus (to produce component C). The events at the nuclear membrane may involve the nuclear pores, where it has been suggested that a normal ATP-dependent process for the transport of cellular material is subverted by the virus [94]. Lysosomal enzymes probably are not involved in the uncoating sequence of adenoviruses [57].

Nucleocapsids of herpesvirus (component A) may also lose their dense cores (component B) at or near the nuclear membrane [452, 458, 753]. In contrast, the papovaviruses may enter the nucleus intact [25, 305] and it has been suggested that they may either pass directly through the nuclear pores [94], or through the nuclear membrane by a more complicated process [305].

The removal of the membrane of the enveloped viruses is the most important step in their uncoating and the resulting nucleocapsids are often able to participate in early macromolecular synthesis without complete disassembly. The envelope of vaccinia virus may be removed during fusion with the membrane of a vesicle or of the cell surface [233] or, alternatively, it may be lost in a pinocytotic vesicle [137]. The core released into the cytoplasm (component A) may require minor alterations or uncoating before it begins messenger RNA synthesis but complete uncoating occurs only later and in the cytoplasm (see below). Some of these events may require participation of the endoplasmic reticulum with which various subviral components of vaccinia were found to be associated [134, 135]. After envelopes of orthomyxo- or paramyxoviruses have been removed at the cell membrane or in a membrane vesicle (sect. V), the helical nucleocapsids (component A) are able to carry out the initial steps of transcription without complete uncoating (see below). The orthomyxoviruses may require entry into the nucleus to account for the accumulation there of antigens [671] but it is also entirely possible that this accumulation is secondary to both transcription and translation.

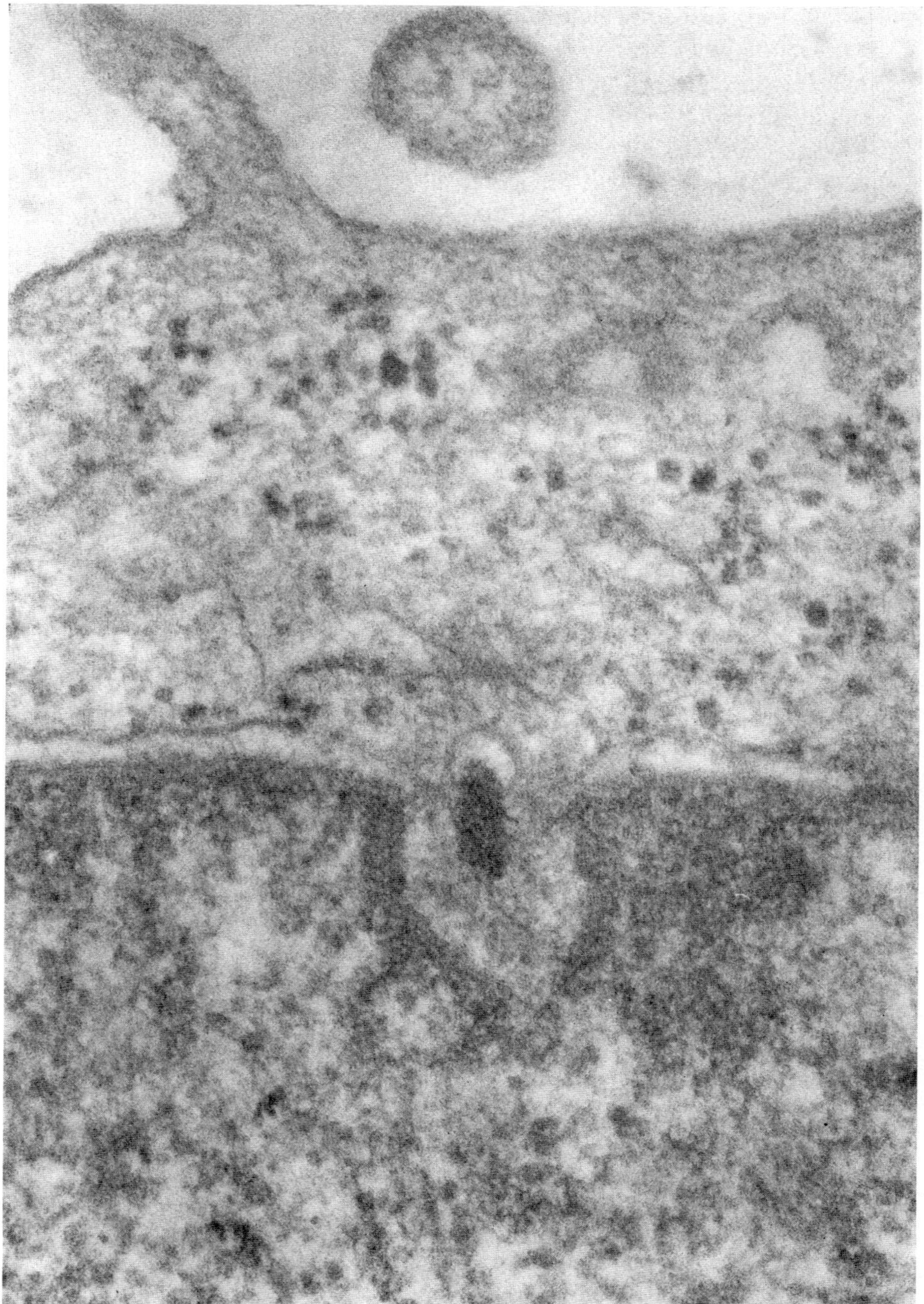

In conclusion, it appears that both the plasma membrane and the nuclear membrane and possibly also lysosomes and the endoplasmic reticulum, are associated with several specific events in virus uncoating. The clarification of their contributions to these steps will require more profound insight into the biochemistry of cellular membranes.

### *B. Subviral Products of* in vivo *and* in vitro *Uncoating of Nonenveloped Viruses*

Only examples of the structural alterations of virions which lead to uncoating will be discussed, particularly those for which there are *in vitro* methods to produce particles or components which resemble those found *in vivo*.

#### 1. Adenoviruses

Adenoviruses undergo three steps of uncoating early in infection. The first of these leads to a particle whose DNA can be solubilized to 70 % by DNase and which is rapidly produced after virions attach to the cell [401]. This particle is designated A in table IV; it is found free in the cytoplasm [401, 459] and cannot be resolved from unaltered virions by sucrose gradient centrifugation but is resolved by isopycnic equilibrium centrifugation in CsCl [401, 656]. There is immunochemical evidence that the penton-fiber proteins have been partly or entirely removed [92, 656] and there is also electron microscopic evidence that these particles have undergone a structural modification [459]; similar particles have been detected in aged preparations of purified virus [656] and methods for producing these from purified virus *in vitro* have been developed [381, 543, 544]. The

---

*Fig. 8.* The core of adenovirus type 7 in the process of being removed from the capsid at the HeLa cell nuclear membrane. The nucleus is at the bottom and a pocket of dense chromatin has formed at the inner margin of the nuclear membrane. The capsid has ruptured and the core is passing into the pocket. Other evidence has also suggested that the core passes from the pocket into the nuclear matrix. From MORGAN *et al.* [459]. ×160,000.

*in vitro* methods release penton-fiber proteins and additional minor components from the 12 vertices and these can be separated from the subviral particles by centrifugation. The loss of the vertex capsomeres probably does not produce antigenic alteration of the remaining hexons. Prolonged treatment, or heating, results in the additional loss of the peripentonal hexons from the capsid [543, 581].

The second step in the uncoating of adenovirus by cells results in the removal of the DNA-protein core from the hexon-capsid. With adenovirus type 2, the core and the capsid sediment together at about half the rate of intact virions but they can be separated by buoyant density centrifugation [401]. The DNA of the core is more sensitive to DNase than is that of the pentonless product of the first uncoating step [401]. An electron micrographic study of this process with adenovirus type 7 has been published [459]; figure 8 is from this report; it shows the uncoating step in progress at the nuclear membrane where the dense core is drawn *en masse* from the capsid shell. Similar results were obtained with adenovirus type 5 [94]. The core has also been freed from the hexon shell *in vitro* by fragmenting the capsid with aqueous solutions of acetone [379], urea [412, 543], or pyridine [543], or by heating in the presence of deoxycholate [579]. A major portion of the 240 virion capsomeres were recovered in geometrically characteristic groups of 9 hexons. Depending upon its concentration, dilute formamide produces large fragments of the capsid or single capsomeres, as well as cores [639]. The cores can also be liberated by the gentle method of freezing and thawing and in this case the hexons are released as single capsomeres [543, 544]. The core produced *in vitro* appears to differ from the core isolated from infected cells in that it aggregates in saline and also has a higher buoyant density in CsCl. These differences may be due to attachment of cellular components to the intracellular cores and it was found that the properties of cores produced *in vitro* by urea treatment resembled those produced *in vivo* when the product was mixed with cell extracts [401].

The third and final step in the uncoating of adenovirus DNA during early infection occurs mainly during the second hour of infection and in the nucleus of the infected cell [401]. The product appears to be DNA largely free of virion protein and it has been isolated as slowly sedimenting fragments after sonic disruption of infected cells, although gentle extraction with alkali reveals that the virion DNA is intact and unnicked [401]. The process whereby the core protein is removed from the DNA is unknown; it seems likely that it requires chemical alteration of the protein.

## 2. Papovaviruses

The subviral components of the papovaviruses have not been studied in great detail, but there is biochemical [25, 508] and structural [305, 342] evidence for cellular uncoating. Intact SV40 virions have been isolated from the nuclei of AGMK cells by CsCl-gradient centrifugation 0.5–2.5 h after infection and these were found to gradually lose protein and to band at a greater density. Similarly, a dense component derived from polyoma virus has also been isolated from infected cells [59]. A *ts* mutant of SV40 which is blocked in an early function of replication, attached, penetrated and was uncoated by host cells in a normal manner, even at the nonpermissive temperature [560]. There has not yet been a close comparison of the *in vivo* and *in vitro* uncoating of papovaviruses. Degradation of SV40 virions in alkaline buffer produces specific products containing 1 or 3 of the internal polypeptides and DNA [299a]. These proteins appear to regulate *in vitro* transcription of the DNA [300] but further work is required to demonstrate that they function *in vivo*.

## 3. Reoviruses

The uncoating of reovirus is incomplete in the infected cell; only a part of the protein of each virion is removed and this loss is ultimately partially replaced so that at the end of infection the parental virions resemble the progeny [91, 616, 620]. The capsids of reovirus appear in the electron microscope to be composed of 2 concentric shells of protein. The inner shell plus the double-stranded RNA constitutes the core; core preparations differ from each other in the extent to which the proteins of the outer shell have been removed (see below).

The nature of the major alterations of the polypeptides of reovirus by the host cell during early infection have been reported [91, 616]. Some of the outer capsid proteins are removed (polypeptide $\sigma 3$) or cleaved (polypeptide $\mu 2$) and the resulting core resembles particles produced from virions *in vitro* by treatment with chymotrypsin [52, 320, 607]. Chymotrypsin can also strip away additional protein from the outer shell under certain conditions of ionic strength and at certain ratios of enzyme to virion [320, 607, 616]. The various chymotrypsin cores differ from each other in buoyant density; those which resemble particles produced during infection show a 0.01–0.02 g/ml greater density than virions while those digested to a maximal extent are about 0.07 g/ml heavier than virions [52, 320].

Reovirus cores produced in the infected cell are functionally equivalent to the cores produced with chymotrypsin *in vitro,* in that they all contain an active RNA polymerase which translates the double-stranded virion genome into messenger RNA segments without the need for further uncoating [52, 608, 616, 650]. Bluetongue virus, which is classified together with the reoviruses, also has a core-like particle which resembles chymotrypsin cores of reovirus by lacking 2 major polypeptides, by having an increased buoyant density and in containing an active polymerase [434]. Other viruses containing double-stranded RNA segments also share structural and functional features with reoviruses and bluetongue virus [395].

The reaction of chymotrypsin with intact reovirus has been dissected into several stages [320]. First the smallest capsid protein $\sigma 3$, which corresponds to 28 % of the virion protein, is fragmented and removed, but this does not alter the buoyant density of the virion in CsCl or the infectivity, indicating that $\sigma 3$ does not play an important role in attachment to the host cell. A second stage commences with the degradation of another polypeptide ($\mu 2$) which results in an increase in buoyant density by 0.02 g/ml, a decrease in infectivity and an activation of the virion polymerase. Ultimately the three surface polypeptides ($\sigma 3$, $\mu 2$ and $\sigma 1$) are completely lost and the product bands at 0.07–0.08 g/ml greater density than virus, possesses maximal polymerase activity and has a specific infectivity 4–5 logs lower than that of virus.

There is evidence that the ionic environment of the virion is important during the uncoating and activation reaction [52, 54, 607]; $K^+$, $Cs^+$ and $Rb^+$ enhance the activation of reovirus polymerase by chymotrypsin while $Na^+$, $Li^+$ and $NH_4^+$ are inactive [54]. The presence of different ions leads to cores of different polypeptide compositions; cores produced in the presence of $Na^+$ may lack polymerase activation and differ in polypeptide composition from cores produced in the presence of $K^+$, but the polymerase of the former may be activated by adding $K^+$ [52]. Also, it has been known for some years that heat alone in the absence of chymotrypsin is sufficient to activate reovirus virion polymerase [53, 608]. The infectivity of reovirus is also both cold-labile and heat-reactivatable in the presence of 2 M $MgCl_2$ [711]. These results suggest that it might be profitable to examine the question of conformational alterations of the reovirus capsid during the activation of virion polymerase. On the other hand, the polymerase activity of cores isolated from infected cells is lost if native soluble $\sigma 3$ polypeptide is added back *in vitro;* a particle is thereby reconstituted which resembles native virions but which is of low infectivity [13].

## 4. Picornaviruses

The subviral components produced during the uncoating of poliovirus, coxsackievirus B3, and rhinovirus type 2 will be discussed together because of similarities of the products of the first step of uncoating. Uncoating of these viruses requires at least two steps, the first of which was recognized by the irreversible alteration of virion infectivity as already discussed in section V. In each of the 3 virus systems, the product of the first step is a particle containing RNA and which sediments at 80–90 % the rate of native virions [127, 200, 399, 400]. This particle lacks the smallest virion polypeptide VP4 [127, 399]. In the case of coxsackievirus B3 [126] and poliovirus [321], the RNA of the noninfective particle was shown to be infective and in these cases the lack of infectivity is probably a result of the inability of the particles to attach to host cells [127, 321]. We shall refer to irreversibly altered particles with these properties as 'A particles' in the following discussion. In the case of rhinovirus type 2, the properties of the cell-associated A particles and of the eluted A particles have been compared in detail with those of subviral noninfective particles produced *in vitro* by treatment of virions at pH 5, as discussed below [362, 399, 400, 402, 488]. Cell-associated A particles are probably formed during the irreversible alteration of poliovirus and coxsackie B3 virus, and eluted A particles are a concomitant of their formation. Similar particles have recently been found to be the major cell-associated product of the cell-mediated eclipse of poliovirus [Lonberg-Holm and Gosser, in preparation].

Rhinovirus A particles can be produced *in vitro* by treatment of virions at pH 5 [362, 402, 488]. Most work has been done with rhinovirus type 2, but A particles can also be produced *in vitro* from other serotypes [488] and the formation of A particles from types 1A, 1B and 14 has also been detected in the cell [399; Lonberg-Holm, unpublished]. A particles produced *in vitro* are also unable to attach to cells [488] and this is sufficient reason for their lack of infectivity. Noninfective subviral particles which contain RNA and lack VP4 have also been produced *in vitro* from poliovirus by treatment at pH 10 [333] or by heating to 50 °C [63]. In the latter case, the particles were shown to sediment at only about 50 % of the rate of intact virions and hence do not resemble A particles produced by cells.

It is possible, however, that faster-sedimenting particles can be obtained by more gentle means [BREINDL, personal commun.].

The formation of noninfective particles, which are unable to attach to host cells, from poliovirus [63, 334] or rhinovirus [402] coincides with an alteration of the antigenic determinants of the surface of the native virions. The determinants of the native virions are referred to as D or N determinants, and the altered set are termed C or H determinants. The widely used D and C designations were derived from bands observed in sucrose gradients of extracts of poliovirus-infected cells [439] and we use them in this discussion. (N and H refer to the determinants of native or heated virions.) There is also some evidence for occasional heterotypic cross-reactivity between the C antigens of various picornaviruses, usually within a virus family [205, 402]. With poliovirus, the conversion of D- to C-reactive particles corresponds in every case but one to a loss of VP4 [63, 65, 333, 334]. The exception is that VP4 is not lost upon antigenic conversion and inactivation of poliovirus with diethylpyrocarbonate [65]. It has thus been proposed that VP4 plays a unique role in both the D reactivity and the ability of poliovirus to attach to host cells [64, 65]. Other evidence that VP4 is D-reactive while located in or on the virion has also been proposed [65], but it is also commonly agreed that VP4, after release from the virus, does not carry D reactivity.

Although this view of VP4 as a protein with the specific function of attaching to the receptors of host cells is an appealing one, we prefer the older or 'classical' view [275], namely that the entire surface of the virion particle plays a role in its antigenicity and ability or lack of ability to attach to host cells. This follows in part from the observation that most of the D-reactive determinants are lost during the antigenic conversion of picornaviruses to C-reactive particles [219, 402, 590] – an alteration too extensive to be ascribed to the simple loss of VP4, the smallest of the virion polypeptides. Secondly, all of the natural top-component particles of rhinovirus type 2 are deficient in VP4 but a portion of these can attach normally to host cells [488] and are D-reactive [402; KORANT *et al.*, in preparation]. The D reactivity of the active portion of the natural top-component particles can also be altered by treatment at pH 5.7 in a manner which resembles the antigenic alteration of native virions at pH 5, and the alteration is blocked by a specific inhibitor of the formation of A particles [400]. Since the attaching and nonattaching natural top-component particles both appear to lack VP4 (but contain VP0 which is probably the uncleaved precursor of VP4 as with poliovirus [312]), the presence or absence of VP4 alone may not explain their antigenic and biological properties. It must be noted, however, that it is not possible to detect the presence of a few molecules of VP4 per particle, and it is possible but unlikely that the presence or absence of such a minor component might determine the biological activity of natural top component. The natural top component of echovirus 12 has also been reported to contain D-reactive antigenic determinants [245].

In the case of poliovirus, the conversion of virions to A particles not only alters their antigenicity but also results in a particle which has enhanced susceptibility to proteolytic enzymes, as shown with 'eluted' virus [200, 275]. This enhanced susceptibility has provided the basis for the 'uncoating' of poliovirus *in vitro*, i. e. uncoating achieved by trypsin treatment of particles which have undergone irreversible alteration *in vitro* [90]. The RNA of intact A particles of poliovirus appears to remain resistant to ribonuclease digestion; in the case of rhinovirus type 2 the RNA is also largely resistant [399] although under some conditions it is susceptible to less than 20% [LONBERG-HOLM, unpublished]. The A particles of polioviruses [321], of coxsackie B3 virus [127], and of the human rhinoviruses [399] are unstable in concentrated solutions of CsCl.

The A particles of picornaviruses can then be viewed as a first unstable intermediate in the process of uncoating. They are probably on the pathway leading to infection because their formation, in the case of rhinovirus type 2, is blocked by 30 μg/ml SDS which specifically also blocks plaque formation [400].

The second step in the uncoating of poliovirus and rhinovirus during infection is not as well established as the first. In the case of the rhinoviruses [400] and poliovirus type 2 [LONBERG-HOLM, in preparation] some A particles are converted in the infected cell to empty capsids (B particles) which sediment at about half the rate of native virions. Other slowly sedimenting capsid products are also formed during the same period and thus the B particles may not be part of the infective pathway and either a relatively intact shell or a fragmented shell may be removed from the RNA. Rhinovirus B particles formed *in vivo* resemble B particles produced *in vitro* by acid [399], heat, or urea treatment [400, 402]. The particles made *in vitro* have been shown to be C-antigenic [402].

The thiopyridine, S-7, inhibits poliovirus infectivity but has been reported not to block the cell-mediated eclipse of infectivity, yet to prevent the subsequent inhibition of host cell protein synthesis which results from an early event in infection [372]. If this is correct, it is possible that S-7 inhibits either the uncoating of, or transport of, poliovirus A particles; the data can, however, also be interpreted to indicate that S-7 produces a partial inhibition of cell-mediated eclipse.

Poliovirus grown in the presence of acridine orange or neutral red dye is sensitive to photodynamic inactivation and this property is usually lost between 15 and 60 min after the virion enters the host cell, as determined by the resistance of infective centers to light [589a, 734]. In the same way, dye-free virus becomes sensitive to photodynamic inactivation after it infects cells containing dye [589a]. The detailed kinetics of the release of infective centers from light-sensitivity after infection with poliovirus containing neutral red indicate that virus first penetrates the cell to a state which is resistant to neutralization by the addition of extracellular neutralizing antibodies and is later uncoated with complex kinetics [423]. A second more effective phase of uncoating occurred after an initial period of uncoating lasting about 20 min at 37 °C or 40 min at 32 °C. Both the first and second stages of uncoating had high energies of activation (about 70 kcal). There are several possible explanations for the existence of a second stage in the kinetics of 'uncoating' and one is that the second stage corresponds to the disassembly of A particles. This would require that neutral red A particles retain a partial sensitivity to photodynamic inactivation, a possibility which has not yet been tested.

The measurement of light-resistant infective centers is a powerful tool in the case of the picornaviruses because only those virions which are successful in initiating infection are scored and those virions which enter abortive pathways do not obscure the final result. In contrast, the measurement of the uncoating of RNA-labeled poliovirus [321] or rhinovirus [399] by the detection of ribonuclease-sensitive intracellular RNA has yielded little useful information because after about one hour most of the cell-associated radioactivity is in a degraded acid-soluble form. This implies that the step in infection subsequent to the formation of A particles is often abortive and that 'successful' uncoating may be one of the major obstacles in the infection pathway.

Studies of the uncoating of the cardioviruses (EMC, mengo, ME, etc.) have not been fruitful. This is often due to slow or inefficient 'eclipse' of infecting virus; the persistence of a significant portion of the infecting virus [229] during the course of infection obscures biochemical investigation of the virus in the infective pathway [286]. The uncoating of three plaque variants of mengovirus has been examined by the reisolation of intact virions from cells at various times after infection [414]. Unfortunately, this method did not include the detergent lysis of virus-receptor complexes nor did it account for radioactivity which was not recovered as intact virus; thus the nature of the products of uncoating remain obscure. The results did show, however, that the rate of loss of intact virus depends upon the plaque variant employed.

Evidence has been provided that the product of uncoating of ME virus may be an empty capsid which is subsequently disrupted to 14S fragments [244]. This process occurs on the cell's surface since most of the RNA is released to the medium and, subsequently, there is a slow release of 14S capsid subunits. Elution of RNA is apparently an abortive reaction since it is decreased by alteration of the conditions of attachment which increases the efficiency of infective center formation. It is noteworthy that 14S capsid subunits can be generated from purified cardiovirus by treatment at about pH 5.7 and in the presence of 0.1 M chloride ion [175, 413, 415]. The fragments contain equimolar quantities of each of the 3 larger virion polypeptides and appear to contain 5 molecules of each while the smallest polypeptide (which may be analogous to VP4 of poliovirus and rhinovirus) is lost [175]. This appears to be somewhat similar to the generation by acid of A and B particles of rhinovirus, which also are composed of the 3 larger polypeptides but which probably contain all 60 molecules of each.

Foot and mouth disease virions are also labile in weak acid and are also disrupted to 12S fragments consisting of equimolar ratios of the three larger virion polypeptides; in this case, however, there is evidence that 3 molecules of each are present in each fragment [75, 667]. These 12S fragments are produced during uncoating of virions by cells [67, 118, 575; BACHRACH and BROWN, private commun.] and have C antigenicity [89]. Empty capsids, which sediment at about 75S, are also detected in infected cells [234, 536] and, as in the case of the cardiovirus, there remains some uncertainty about the primary uncoating step under conditions leading to an infection. It was recently found that the smallest polypeptide (VP4) of FMDV can be recovered as an aggregate after virions are disrupted by treatment at pH 6.5 and this aggregate has C antigenicity [668]. Other evidence, based upon labeling with radioactive iodine, suggests that VP4 is located inside the FMDV capsid before acid disruption [668].

## 5. A Bacteriophage System

Although detailed discussion of bacteriophage-cell interaction is outside the scope of this review, it is instructive to compare some aspects of the early interactions of picornaviruses or other nonenveloped viruses with the eclipse and uncoating of the small DNA-phage ØX174. After ØX174 attaches to cells, it undergoes rapid and temperature-dependent eclipse of infectivity (with an activation energy of 37 kcal/mol) [480]. The product is unaltered in protein composition but appears to have undergone a conformational alteration since it sediments more slowly than infective phage; a variable portion of the DNA also protrudes from the particle [479]. A similar *in vitro* alteration can be induced by treatment with 0.1 M $CaCl_2$ [308]. Virus mutants are also known with altered ability to undergo cell-mediated eclipse and *in vitro* heat inactivation [162].

The uncoating of ØX174 appears to occur at the cell surface and the penetration of the DNA can be blocked by starvation of the host [353, 479]. Uneclipsed phage which has attached to the host cell at low temperature cannot be neutralized by incubation with neutralizing antisera [479]. In this case there can be no doubt that the uneclipsed phage is located outside the cell wall and on this basis there is reason to reserve final conclusions in those studies with animal viruses which rely upon the escape from sensitivity to antibody neutralization as a criterion for penetration or pinocytotic uptake.

## *C. Subviral Products of* in vivo *and* in vitro *Uncoating of Enveloped Viruses*

### 1. Poxviruses

Virions of vaccinia are very large and have a correspondingly complex structure [724]. There are at least 30 virion polypeptides [586]; two of these contain hexosamine [586] and are presumed to be located outside of the core but within the envelope because they are removed by detergent [282] but are not accessible at the virion surface by the criteria of iodination or enzymatic digestion of intact virus [586].

The components found during the uncoating of poxviruses have been reviewed [319, 739]. Early work [138, 316, 317, 585, 693] showed that poxviruses are uncoated in two steps in the infected cell. In the first step, the virion envelope is removed from the 'core'. DNA is released from the core in the second step, which occurs after a lag and is dependent upon RNA and protein synthesis. The lag was eliminated in cells infected with a poxvirus prior to the infection with labeled virions. Initially it was thought that both uncoating steps were required for viral DNA to assume its genetic function [317] and hence that complete uncoating required the induction and functioning of a host cell enzyme. However, it was found that poxvirus messenger RNA could be transcribed in the absence of the complete uncoating of the virion DNA and that the virions themselves contained RNA polymerase [739]. Subsequent work has confirmed that cores with active polymerase can be isolated from the cytoplasm of the infected cell [135].

At least 5 subviral vaccinia components have been separated from infected cells on the basis of their buoyant density and some of these appear to be complexed with cell membranes [134, 135, 283, 587]. One of these lacks DNA and may correspond to empty capsids, while a slowly sedimenting component which contains mainly DNA and which is largely but not completely sensitive to deoxyribonuclease may correspond to the virus 'chromosome' [283, 587] and corresponds to component B of table IV. A minor component resembles intact virions and contains 'lateral bodies' while a major component corresponds to cores devoid of the virion envelope and lateral bodies. Subviral cores have also been produced *in vitro* by a variety of methods [179, 283, 284, 447]. A close comparison of their sedimentation rates shows that particles produced *in vitro* which contain lateral bodies and cores both have *in vivo* counterparts [283], but that in the case of the cores, the polypeptide patterns are not entirely identical [587]. It is not known if cores undergo conformational alteration in the activation of their RNA polymerase nor is it known by what mechanism the virion 'chromosome' is released, except that a newly synthesized protein is required.

## 2. Herpesviruses

The architecture of the herpes virion and of some subviral components has been reviewed [568]. The nucleoprotein (capsid plus core) is enclosed in the virion's glycolipoprotein envelope. Recent work indicates that the DNA in the core is 'spooled' up into a toroidal mass about a cylindrical structure which is presumed to be protein [213]. A detailed model for the rest of the virion has a 2-layered envelope and a 3-layered capsid surrounding the core, and subviral particles corresponding to each possible level of uncoating may have been isolated [569]. Some of these particles were isolated from cells late in infection and are intermediates in virion assembly [220] and hence can be expected to differ from the intermediates in uncoating. The location of some of the approximately 22 virion polypeptides within these subviral particles is known [263, 505] but much biochemical work remains. The ultrastructural studies of early infection have provided evidence for several steps in the uncoating of herpesviruses, but there is still disagreement on the location of the first step in which the virion envelope is removed (sect. V). It is thought that the core DNA is removed from the capsid at a location close to the nucleus (sect. VI.A).

### 3. Oncornaviruses

The oncorna-, or leukoviruses, also have a complex structure and contain 8 or 9 polypeptides [48]. All of the glycoproteins are exposed on the surface of the virion envelope [554] and appear to contain all or most of the type-specific surface antigens [28, 49, 169]. Removal of the outer envelope by detergent and at low ionic strength produces 'cores' which can be purified by gradient centrifugation [20, 104, 549]; these may have a complex structure which can be resolved into a core 'shell' and a filamentous inner ribonucleoprotein [50, 473]. The intact core is unstable at physiological salt concentration and hence its RNA would appear to be uncoated if it were isolated from an infected cell [20]. Rous sarcoma virus cores contain a major protein which probably coats and interacts with the virion RNA [50, 549] to form the internal ribonucleoprotein thread [403, 473, 585]. This protein is a group-reactive antigen and it even retains antigenicity after phenol extraction and SDS-gel electrophoresis [28]. Unidentified proteins in the core have enzymatic activity; the RNA-directed DNA polymerase of Temin and Baltimore [674] has recently received much attention and has also, together with other virion-associated enzymes, been reviewed by these authors. At the present time there is very little biochemical information upon which to relate oncornavirus virion structural components to intermediates in *in vivo* uncoating.

### 4. Orthomyxoviruses and Paramyxoviruses

The structures of representative orthomyxo- and paramyxovirus virions have been discussed in section II.A and the removal of their envelopes in the first step of cellular uncoating has been discussed in section V.

The orthomyxoviruses (influenza and fowl plague) are somewhat smaller than the paramyxoviruses and contain a smaller and more flexible internal ribonucleoprotein [111] which may be 'looped' in regions into thicker and distinctly double-stranded segments [541, 603]. The genome, of $4–5 \times 10^6$ molecular weight, resides in six or seven single-stranded RNA segments rather than in an intact molecule, and the biological significance of the segmentation of the influenza virus genome has been discussed in a recent review [650]. The membrane components can be separated from the internal ribonucleoprotein after lysis with ether or detergent [627]; a particularly useful method has been described by PONS [540].

The internal proteins of influenza virions include the major nucleocapsid protein, and at least two other polypeptides which may have enzymatic functions [43, 344, 627]. Myxovirus virions have been found to contain an RNA-dependent RNA polymerase [41, 98] which is associated with the ribonucleoprotein strands and transcribes all of the genome segments [43]. The RNA-polymerase of fowl-plague-virus-infected cells [600] as well as of Sendai virus virions is inhibited by polyvinyl sulfate. Polyvinyl sulfate competes with virion RNA for the binding of the influenza nucleocapsid protein; this affords a simple system for *in vitro* uncoating [227]. Capsid protein might interfere with transcription but it is also possible that capsid protein enters into the transcription complex together with RNA and a polymerase, in which case the genome need never be fully 'uncoated'.

The nucleocapsid strands of the paramyxoviruses resemble strands which accumulate in the infected cell during virion assembly [110, 201]. Within the virion envelope and presumably in association with the ribonucleoprotein is an RNA polymerase which transcribes short, probably monocistronic virus messenger RNA molecules early in infection [296, 562, 563, 647]. A requirement for this enzyme in the infective process may explain why purified polycistronic virion RNA of the paramyxoviruses is not infective [650]. The polypeptides of the transcriptase are not established, nor is it known whether the virion RNA is uncoated when it functions as a template. Various polyanions stimulate or inhibit transcriptase activity of Sendai virions and they may act by modifying the extent or nature of the binding of capsid protein to RNA, although some evidence argues against these possibilities [646]. The nucleocapsid RNA is not sensitive to solubilization by ribonuclease but uncoating by this criterion has been reported to occur in infected cells and also to be inhibited by puromycin [177]; these observations have not yet been confirmed and the possibility of structural modification of the parental nucleocapsid during early infection requires further examination.

## 5. Rhabdoviruses

The distinctive bullet shape of the rhabdoviruses belies structural similarities to the paramyxoviruses [111]. The rhabdoviruses resemble the larger paramyxoviruses in having a single-stranded nonsegmented RNA [650] and a virion RNA-dependent RNA polymerase with similar properties [1, 22, 42, 296]. The major protein of the virion core nucleocapsid is a group-specific antigen [327]; the core can be released in either a coiled or uncoiled form after treatment of virions with various detergents [80, 81]. Treatment with a neutral detergent and high salt concentrations produces an unraveled nucleoprotein complex which can be purified and which contains 3 polypeptides and both transcriptase activity and infectivity if tested in the presence of DEAE-dextran [663]. A similar complex may be released into the infected cell. The uncoiled complex sediments at about half the rate of intact virus but a recent attempt to identify such a component in cells infected with labeled VSV failed, and the virion RNA was rapidly converted to slowly sedimenting, acid-insoluble material [743]. The ribonucleoprotein might have been fragmented during disruption of the cells, or the slowly sedimenting product might have been 'uncoated' RNA; unfortunately, the slowly sedimenting product was not tested for ribonuclease sensitivity. The virion transcriptase is known to function for long periods within the infected cell, even in the presence of cycloheximide. Cells treated with interferon before infection are less able to support this virus-specific RNA synthesis [427, 432].

# VII. Host Cell Macromolecular Synthesis during Early Interaction

## *A. Introduction*

The study of macromolecular synthesis after viral infection usually requires the synchrony obtained by infecting with high multiplicities [115]. At high multiplicities of infection, many artifacts may interfere with the study of host macromolecular synthesis:

(a) Enveloped RNA viruses like mumps and NDV are toxic at high multiplicities [264, 545, 574, 697].

(b) Preparations of rhabdoviruses [116, 298, 299], myxoviruses [599, 696], picornaviruses [107], papovaviruses [686] and probably other viruses may contain incomplete interfering particles which can alter the course of infection in each cell with which they interact.

(c) Since the ratio of particles to plaque-forming units is 10–1,000 for most animal viruses, the particles which do not replicate may have a profound effect on host metabolism.

(d) Impurities in the virus used for infection may influence the metabolic events in the host cell, for example, structural subunits of the virions may affect the course of macromolecular synthesis [391].

(e) Finally, if mycoplasma are present in the viral inoculum, they may affect both cell growth and metabolism [202, 471, 539].

It is also important to consider that the incorporation of radioactive precursors into DNA, RNA or protein may be affected by changes in cellular permeability and variations in intracellular pool sizes caused by the infection. Furthermore, it is necessary to differentiate between host and viral metabolism. This may be achieved, in some cases, by separating infected cells into nuclei and cytoplasm, as in the case of vaccinia where viral replication occurs only in the cytoplasm. Differences in the properties of viral and cellular macromolecules may also be utilized as with herpes- and adenoviruses where viral DNA may be resolved from cellular DNA by buoyant density centrifugation. Sometimes it is also feasible to selectively inhibit the host cell macromolecular processes to assess the rate of viral synthesis, for example, actinomycin D has been used to study the replication of several RNA viruses.

With these precautions in mind, it appears that inhibition or stimulation of macromolecular synthesis in infected cells depends on both the virus and the cell. There is no regular pattern and no evidence that the synthesis of one macromolecule is more readily inhibited than that of another. It is also apparent that the *inhibition* of macromolecular synthesis is most readily observed in continuous cell lines and suspension cultures; this is mainly due to the fact that macromolecular synthesis and especially DNA synthesis is arrested in confluent primary monolayer cultures. Viruses like picornaviruses and herpesviruses may inhibit all host macromolecular synthesis [272, 567, 660, 756], but NDV appears to primarily affect protein synthesis and only a few strains inhibit host RNA synthesis [725, 733].

Drastic inhibition of host DNA, RNA and protein synthesis has many times been thought to result from an interdependence between the synthesis of one macromolecule and another. The picornavirus system has furnished evidence that inhibition of host protein synthesis precedes that of host RNA synthesis [15, 441, 756] and also that the rate of inhibition of protein synthesis is more rapid than the decay of host messenger RNA. Furthermore, in poliovirus-infected cells, host polyribosomes appear to disaggregate more rapidly than in cells treated with actinomycin D [143, 515]. Host DNA synthesis is inhibited last in picornavirus infection, but synthesis of all three macromolecules is inhibited to the same extent. There is, however, distinct host dependence on the inhibitory processes as exemplified by mengovirus which grows equally well in L and HeLa cells. In L cells, the inhibition of host RNA synthesis is rapid and occurs soon after infection [207, 441], but in HeLa cells it is late and less severe [441]. The rate of inhibition of RNA and protein synthesis by mengovirus in HeLa cells parallels that observed for poliovirus [441, 756]. Furthermore, the time and extent of inhibition of host RNA and protein synthesis by mengovirus differed in two strains of Novikoff hepatoma cells propagated *in vitro* [535].

The most sensitive step in host RNA synthesis appears to be the synthesis of ribosomal precursor RNA [149, 702]. The synthesis of 45S ribosomal precursor RNA as well as its conversion in the nucleus to 32S precursor and 18S ribosomal RNA were depressed as was the assembly of the large ribosomal subunit in picornavirus infection [149], while the 45S precursor was degraded in herpesvirus infection [702]. It is worth pointing out that, in uninfected mammalian cells, these processes are very sensitive to inhibition of protein synthesis or to amino acid starvation [410, 692]. The synthesis of mRNA or tRNA is less affected during viral infection.

The inhibition of host cell macromolecular synthesis can occur either late in the infectious cycle, as for adenovirus [221] and paramyxoviruses [733], or early as for some picornaviruses [441], herpes [566] and vaccinia [325, 463]. We will only consider here the early effect on macromolecular synthesis occurring before the replication of virion protein and nucleic acid. Inhibition of macromolecular synthesis has been divided into two classes depending on whether it is mediated by virion components or by virus-induced products. A stimulation of macromolecular synthesis by some of the DNA tumor viruses has been observed mainly in monolayer cultures, and this will also be discussed.

### *B. Inhibition by Virion Components*

*Vaccinia virus* inhibits DNA and protein synthesis of host cells early in infection [325, 463], but host RNA synthesis is inhibited only after 3–4 h [31, 584]. Inhibition of HeLa cell DNA synthesis early in infection by heated or UV-light-irradiated virus is similar to that obtained with untreated virus, and treatment of infected cells with fluorodeoxyuridine prevents the synthesis of viral DNA, but does not diminish the inhibition of cellular DNA synthesis [325]. UV-light-inactivated purified vaccinia virus also inhibits host protein synthesis, but heated or detergent-treated virus is ineffective [463]. The inhibition of host protein synthesis is multiplicity-dependent and is unaffected by actinomycin D [193, 606] or interferon [322]. It appears, therefore, safe to conclude that the inhibitions of host DNA and protein synthesis are probably brought about by a constituent of vaccinia virions. Recently Pogo and Dales [538a] proposed that a component of the invading virus is responsible for the inhibition of host DNA polymerase early in vaccinia virus infection. They also speculate that the host cell nuclear DNA polymerase is inhibited as a consequence of hydrolysis of nascent single-stranded DNA molecules by a virion enzyme. The inhibition of host protein synthesis, however, also depends on the host, since it appears less pronounced in chick cells than in L cells [JUNGWIRTH, personal commun.].

*Adenovirus.* Although inhibition of the macromolecular synthesis of the host cell in adenovirus is a late event occurring after onset of viral DNA and protein synthesis [221], it has been claimed that a structural component of the virion mediates the inhibition. LEVINE and GINSBERG [391, 392] reported that the fiber antigen combines irreversibly with DNA of KB cells and blocks the synthesis of host DNA and protein within 24 h. It is, however, not conclusively shown that the fiber protein is responsible for this, since other viral-induced proteins copurify with the fiber under the conditions they used [655].

*VSV virus.* It has also been reported that UV-irradiated complete or defective vesicular stomatitis virus has an undiminished capacity to inhibit host RNA synthesis [299, 704], again suggesting that a component of the virion mediates inhibition. Diminished synthesis of host protein was also induced by VSV, even in the presence of interferon [742], but subsequent studies suggest that there is both a multiplicity-dependent UV-insensitive inhibition and a progressive UV-sensitive inhibition of protein synthesis [723].

### *C. Inhibition by Virus-Induced Products*

*Picornavirus.* Most evidence on the inhibition of synthesis of host macromolecules by viruses has been obtained with picornaviruses. Poliovirus inhibits cell RNA and protein synthesis very shortly after infection, even in the presence of guanidine [15, 516] or at 42 °C [199], conditions which prevent detectable viral RNA synthesis. The viral inhibition of host RNA synthesis was prevented or diminished by the addition of puromycin [149, 279] or cycloheximide [730] or by the addition of the amino acid analogue parafluorophenylalanine [21] during and after infection. A sparing effect on inhibition of protein synthesis by cycloheximide has also been revealed in a reversion experiment [730]. Furthermore, UV-inactivation of the virus prevented inhibition [207, 516]. These data were interpreted to mean that inhibition requires protein synthesis early in infection before viral RNA was replicated. It has recently been shown that poliovirus defective particles, deleted in genes for structural proteins, are able to inhibit protein and RNA synthesis of the host cell [106].

*Herpesvirus* also inhibits host macromolecular synthesis early in infection and before the majority of the viral proteins are made [566]. Although conflicting reports [481, 660] have been presented concerning the role of virion constituents in host metabolism, several studies suggest that a virus-induced product is responsible [35, 660]. Puromycin decreased the rate of inhibition of host DNA synthesis in rabbit kidney cells infected with pseudorabies [35]. UV-irradiated herpes simplex virus has also lost its capacity to disaggregate Hep-2 cell polyribosomes and to inhibit protein synthesis [660].

*Myxoviruses and arboviruses.* Enveloped viruses such as myxoviruses and arboviruses extensively inhibit host protein and RNA synthesis after infection. The inhibition occurs, however, 5–9 h after infection. On the other hand, some paramyxoviruses like SV5 only rarely inhibit host macromolecular synthesis [281].

## *D. Mechanism of Inhibition*

### 1. DNA Synthesis

It appears established that inhibition of host DNA synthesis is neither the consequence of a degradation of host DNA as for bacterial viruses, nor of competition of virus and host for enzymes and precursors [35, 345]. One possibility is the mechanism suggested for adenovirus [392] where a viral protein binds to host DNA and thereby inhibits the DNA polymerase of the cell. A second alternative is that the inhibition of cell DNA synthesis is a consequence of the inhibition of cell protein synthesis at some stage prior to S phase. Complete inhibition of DNA synthesis may take at least 10 h in an asynchronous culture in which cells divide every 18 h and interference with protein synthesis in earlier periods of the cell cycle may therefore give a late inhibition of DNA synthesis. It has recently been demonstrated that the proteins synthesized during the Gl phase are primarily inhibited during infection of L cells by meningopneumonitis agent [398]. It has also been reported that the oligo nucleotide poly-A of reovirus may be the mediator of inhibition of DNA synthesis [249].

## 2. RNA Synthesis

The inhibition of host RNA synthesis observed early in picornavirus and other virus infections is not an artifact since it has been shown that host RNA has the same lifetime in uninfected and infected cells [207, 435] and also since the rate of entry and pool sizes of nucleosides and nucleotides, respectively, are the same in uninfected and infected cells [435, 436]. However, the primary target appears to be ribosomal RNA synthesis [149, 702] and this may again be a secondary effect of the inhibition of protein synthesis, since ribosomal RNA synthesis is generally sensitive to inhibitors of protein synthesis as well as to amino acid starvation [410, 692].

## 3. Protein Synthesis

Inhibition of protein synthesis may be due to disaggregation of polyribosomes [143, 514, 515, 659]. The mechanism for this disaggregation is unclear, but WILLEMS and PENMAN [730] have demonstrated that it is prevented in poliovirus-infected cells treated with cycloheximide, in which case the host polyribosomes remain intact with regard both to number and to size. When the drug was removed the polyribosomes decreased in number but not in size. Available evidence also excludes random degradation of host messenger RNA and the appearance of defective ribosomes, which cannot participate in translation [341, 652]. It is, however, noteworthy that high concentrations of inhibitors were needed to prevent the inhibition of host protein synthesis by virus. Also, the inhibition of host macromolecular synthesis in abortively infected cells appeared to be more sensitive to the drug than in productively infected cells. Furthermore, host protein synthesis was also inhibited in infected cells treated with interferon at a concentration sufficient to prevent virus multiplication. These results suggest that the inhibitor of protein synthesis probably affects initiation and that it functions catalytically with high efficiency during productive infection [385]. Even trace amounts of this substance made in presence of chemical inhibitors or interferon would then be sufficient to inhibit host protein synthesis. Double-stranded RNA at low concentration inhibits host protein synthesis both *in vivo* and *in vitro* [117, 182]; and it may exert this effect by inactivating a host molecule needed in initiation [147]. It should be noted that the synthesis of viral double-stranded RNA in early picornavirus infection can occur only after the synthesis of some viral protein.

## *E. Stimulation of Macromolecular Synthesis*

The papovaviruses, polyoma and SV40, have been shown to induce synthesis of cellular DNA during productive infection before or at the onset of viral DNA replication. This effect is only observed in cells which have stopped multiplying because of the depletion of growth factors in their medium or because they have become confluent [174, 209, 685]. A similar stimulation of primary monolayer cultures by adenoviruses has been reported [384], and it is not discerned in actively growing suspension cultures [268]. DULBECCO *et al.* [172] showed that stationary mouse kidney cultures synthesized about 10-fold more DNA after infection with polyoma virus and that most of this is cellular. This stimulation of DNA synthesis was accompanied by the induction of several enzymes involved in DNA metabolism. Since most resting cells are arrested at the same stage, the post-mitotic Gl phase of the growth cycle [685, 722], virus infection may release the block to the S phase. The papovaviruses are able to induce DNA synthesis in productively infected cells even in the presence of low concentrations of 5-fluorodeoxyuridine [36]. Most of the newly synthesized DNA contains one parental strand made before infection [330, 721] and the cellular DNA synthesized after infection is the same as that isolated from uninfected cells. Furthermore, the synthesis of histones appears to be induced concomitantly with induction of cellular DNA synthesis [611, 735]. These results indicate that DNA synthesis is induced in all cellular DNA and that replication of the cellular genome is accompanied by synthesis of protein components associated with chromatin. The enhancement of cellular DNA synthesis during productive infection by SV40 or polyoma virus has been observed in several different types of cells of several different species. There is only one example of a stationary culture of cells which support the lytic cycle of a papovavirus without showing a concomitant increase in cellular DNA synthesis and that is BSCl monkey cells infected with SV40 [218, 557]. These cells do, however, respond to serum factors with increased DNA synthesis as regularly found for topographically inhibited cells.

Stimulation of cellular DNA synthesis appears to depend on part, but not all, of the viral genome, since the capacity of polyoma virus to induce host cell DNA synthesis is inactivated by nitrous acid or ultraviolet irradiation at a rate up to 5 times slower than the inactivation of infectivity [156, 217]. Both rates follow first order kinetics, indicating that only a part of the viral genome is involved in cellular DNA synthesis. Furthermore, one temperature-sensitive mutant of polyoma (ts3) appears to lack the ability to induce cellular DNA synthesis in 3T3 cells at a nonpermissive temperature [171]. In addition, the induction of cellular DNA synthesis precedes late events and is sensitive to interferon [170, 173] and presumably therefore depends on the translation of viral messenger RNA.

Thus, in summary, productive infection with the papovaviruses in particular but also with adenovirus in confluent cells usually induces cellular DNA synthesis. The significance of this induction for replication of viral DNA, as well as the mechanism of induction, is still unknown.

# VIII. Conclusions

This review has emphasized the specificity involved in viral attachment at the plasma membrane as revealed by differences in the receptors for different viruses. There are at least 5 and probably more distinct receptors employed by various picornaviruses, as revealed by competition studies or susceptibility to enzymatic inactivation. In other cases, the same receptors on the cell surface may be recognized by a number of distinctly different viruses; for example, the NANA-containing receptors for myxo-, paramyxo-, cardio-, and polyoma viruses. In cases such as these, there may be partial overlap in that myxoviruses may recognize all polyoma receptors, and others in addition. It should be emphasized that a 'receptor' may consist of a single molecular entity while a 'receptor site' may consist of identical or different molecules located in or on the plasma membrane. Evidence that the attachment of virions to cells requires the participation of many receptors to form a multivalent bond has been discussed in detail.

The virion polypeptides recognized by the plasma membrane have been identified both for adeno- and myxoviruses. For other viruses with cubic symmetry, like picornaviruses, it is not yet established whether the surface pattern of the virion or single polypeptides are recognized. The first alternative is favored by the fact that the picornavirus virion is able to undergo what appears to be a cooperative alteration (D to C antigenic conversion) and hence the entire structure may be viewed as a single protein built of subunits.

The dispute between those who adhere to the viropexis model for virion penetration and others who favor either direct penetration for nonenveloped virions or fusion for enveloped virions has been conducted at times in a rather obtuse style. The personal commitment to one or the other view has engendered detailed investigation which has provided much interesting information but which has not settled the basic dispute. It is evident that the problem cannot be solved without the use of new approaches; these may involve the use of specific inhibitors. We are of the opinion that pinocytosis of virions occurs widely and that it may play a role in the entry of at least some viruses. In many if not all cases, pinocytosis also leads eventually to the entry of virus into lysosomes which may be a host defense mechanism leading to an abortion of the infective process. In other cases, direct penetration of nonenveloped virions may be a concomitant of irreversible alteration; this may occur either from the cell surface or through the membrane of the pinocytotic vesicle. It also seems likely that in many cases, enveloped virions fuse with the cell membrane although this fusion may at times occur on the cell surface and at times with the membrane of the vesicle. Furthermore, the postulation of enzymatic activity leading to the disruption of virion and cell membranes at the external surface of the cell, such as might be involved in the fusion process, is no less probable or reasonable than the evocation of special enzymatic activities leading to the simultaneous lysis of the membranes of vesicles and intravesicular virions.

Viruses fall in two categories with respect to the way they initiate the functioning of their genome in the host cell. In the case of nonenveloped viruses, uncoating constitutes either the loss of capsid protein or an altered relationship between the nucleic acid and the capsid protein, or both. In some instances, loss of capsid protein occurs with the aid of host proteolytic enzymes, as in the case of reoviruses. In others, uncoating is initiated by what appears to be conformational alteration of the capsid which leads to partial loss of capsid protein, as is the case with adenoviruses which lose their vertex capsomeres early in infection. In other cases, uncoating may be initiated by a rearrangement of the virion architecture, as is the case with picornaviruses whose antigens and physical properties are altered during 'eclipse'; the loss of a small virion polypeptide in these cases may possibly reflect rather than cause the major structural alteration. In other instances – so far described only for bacteriophage ØX174 – this conformational alteration may not involve the loss of any of the virus components. After the first step in uncoating, the virion nucleic acid achieves only a *limited* accessibility to the internal environment of the infected cell and complete uncoating occurs later, if at all.

With enveloped viruses, the envelope itself resembles the cell envelope and the problem is formally related to that of producing a heterokaryon bringing two functional genomes into one envelope. After this is achieved, the virion nucleoprotein capsid may begin its function, without need of further rearrangement. Later in infection, possibly as a result of a virus-induced enzymatic activity as in the case of the poxviruses, the genome may be uncoated further. In all cases, with enveloped or nonenveloped viruses, the details of the interaction of host or virion proteins with the virion nucleic acid can be expected to regulate the template activity of the latter. However, since the RNA alone of many animal viruses possesses limited infectivity, this regulation by proteins is not mandatory.

# IX. References


1 Aaslestadt, H. G.; Clark, H. F.; Bishop, D. H. L., and Koprowski, H.: Comparison of the ribonucleic acid polymerases of two rhabdoviruses, Kern Canyon virus and vesicular stomatitis virus. J. Virol. *7:* 726–735 (1971).

2 Abodeely, R. A.; Lawson, L. A., and Randall, C. C.: Morphology and entry of enveloped and deenveloped equine abortion (herpes) virus. J. Virol. *5:* 513–523 (1970).

3 Acheson, N. H. and Tamm, I.: Purification and properties of Semliki Forest virus nucleocapsids. Virology *41:* 306–320 (1970).

4 Ackermann, W. W.; Ishida, N., and Maassab, H. F.: Growth characteristics of influenza virus concerning the binding of virus by host cells. J. exp. Med. *102:* 545–554 (1955).

5 Allen, R.; Finkelstein, R. A., and Sulkin, S. E.: Viral inhibitors in normal animal sera. Texas Rep. Biol. Med. *16:* 391–421 (1958).

6 Allison, A. C.: Observations on the inactivation of viruses by sulfhydryl reagents. Virology *17:* 176–183 (1962).

7 Allison, A. C.; Buckland, F. E., and Andrewes, C. H.: Effects of sulfhydryl reagents on infectivity of some viruses. Virology *17:* 171–175 (1962).

8 Allison, A. C.; Davies, P., and DePetris, S.: Role of contractile microfilaments in macrophage movement and endocytosis. Nature new Biol. *232:* 153–155 (1971).

9 Allison, A. C. and Valentine, R. C.: Virus particle adsorption. II. Adsorption of vaccinia and fowl plague viruses to cells in suspension. Biochim. biophys. Acta *40:* 393–399 (1960).

10 Allison, A. C. and Valentine, R. C.: Virus particle adsorption. III. Adsorption of viruses by cell monolayers and effects of some variables on adsorption. Biochim. biophys. Acta *40:* 400–410 (1960).

11 Almeida, J. D. and Waterson, A. P.: The morphology of virus-antibody interaction. Adv. Virus Res. *15:* 307–338 (1969).

12 Anderer, F. A.: Recent studies on the structure of tobacco mosaic virus. Adv. Protein Chem. *18:* 1–35 (1963).

13 Astell, C.; Silverstein, S. C.; Levin, D. H., and Acs, G.: Regulation of the reovirus RNA transcriptase by a viral capsomere protein. Virology *48:* 648–654 (1972).

14 Axler, D. A. and Crowell, R. L.: Effect of anticellular serum on the attachment of enteroviruses to HeLa cells. J. Virol. *2:* 813–821 (1968).

15 Bablanian, R.; Eggers, H. J., and Tamm, I.: Studies on the mechanism of poliovirus-induced cell damage. I. The relation between poliovirus-induced metabolic and morphological alterations in cultured cells. Virology *26:* 100–113 (1965).

16 Bächi, T.: Electron microscopic investigations on the cycle of infection of influenza A virus in Ehrlich ascites tumor cells. Path. Microbiol. *36*: 81—107 (1970).

17 Bächi, T. and Howe, C.: Fusion of erythrocytes by Sendai virus. A new mechanism. J. Cell Biol. *55:* 10a (1972).

17a Bächi, T. and Howe, C.: Fusion of erythrocytes by Sendai virus studied by electron microscopy. Proc. Soc. exp. Biol. Med. *141:* 141–149 (1972).

18 Bachrach, H. L.: Foot and mouth disease. Annu. Rev. Microbiol. *22:* 201–244 (1968).

19 Bachtold, J. G.; Bubel, H. C., and Gebhardt, L. P.: The primary interaction of poliomyelitis virus with host cells of tissue culture origin. Virology *4:* 582–589 (1957).

20 Bader, J. P.; Brown, N. R., and Bader, A. V.: Characteristics of cores of avian leukosarcoma viruses. Virology *41:* 718–728 (1970).

21 Baltimore, D.; Franklin, R. M., and Callender, J.: Mengovirus-induced inhibition of host ribonucleic acid and protein synthesis. Biochim. biophys. Acta *76:* 425–430 (1963).

22 Baltimore, D.; Huang, A. S., and Stampfer, M.: Ribonucleic acid synthesis of vesicular stomatitis virus. II. An RNA polymerase in the virion. Proc. nat. Acad. Sci., Wash. *66:* 572–576 (1970).

23 Baluda, M. A.: Loss of viral receptors in homologous interference by ultraviolet-irradiated Newcastle disease virus. Virology *7:* 315–327 (1959).

24 Barbanti-Brodano, G.; Possati, L., and La Placa, M.: Effect of phospholipase B (lysolecithinase) on polykaryocytogenic and hemolytic activities of Sendai virus. Arch. vet. ital. *23:* 69–72 (1972).

25 Barbanti-Brodano, G.; Swetly, P., and Koprowski, H.: Early events in the infection of permissive cells with simian virus 40: adsorption, penetration, and uncoating. J. Virol. *6:* 78–86 (1970).

26 Barclay, G. R.; Flewett, T. H.; Keller, E.; Halsall, H. B., and Spragg, S. P.: Effect of polymerized orosomucoid on some strains of influenza virus. Biochem. J. *111:* 353–357 (1969).

27 Baron, S.; Friedman, R. M., and Buckler, C. E.: Properties of poliovirus inhibitor from monkey brain. Proc. Soc. exp. Biol. Med. *113:* 107–110 (1963).

28 Bauer, H. and Bolognesi, D. P.: Polypeptides of avian RNA tumor viruses. II. Serological characterization. Virology *42:* 1113–1126 (1970).

29 Bauer, H. und Wigand, R.: Eigenschaften der Adenovirus-Hämagglutinine. Z. Hyg. InfektKr. *149:* 96–113 (1963).

30 Becht, H.; Rott, R., and Klenk, H. D.: Effect of concanavalin A on cells infected with enveloped RNA viruses. J. gen. Virol. *14:* 1–8 (1972).

31 Becker, Y. and Joklik, W. K.: Messenger RNA in cells infected with vaccinia virus. Proc. nat. Acad. Sci., Wash. *51:* 577–585 (1964).

32 Bengtsson, S.: Mechanism of dextran sulfate inhibition of attenuated poliovirus. Proc. Soc. exp. Biol. Med. *118:* 47–53 (1965).

33 Bengtsson, S.; Dinter, Z., and Philipson, L.: Genetic markers associated with virulence of foot-and-mouth disease virus. Proc. Soc. exp. Biol. Med. *113:* 1019–1022 (1963).

34 BENGTSSON, S.; PHILIPSON, L.; PERSSON, H., and LAURENT, T.: The basis for the interaction between attenuated poliovirus and polyions. Virology *24:* 617–625 (1964).

35 BEN-PORAT, T. and KAPLAN, A. S.: Mechanism of inhibition of cellular DNA synthesis by pseudorabies virus. Virology *25:* 22–29 (1965).

36 BEN-PORAT, T. and KAPLAN, A. S.: Correlation between replication and degradation of cellular DNA in polyoma virus-infected cells. Virology *32:* 457–464 (1967).

37 BERLIN, R. D.: Effect of concanavalin A on phagocytosis. Nature new Biol. *235:* 44–45 (1972).

38 BETEL, I. and VAN DEN BERG, K. J.: Interactions of concanavalin A with rat lymphocytes. Europ. J. Biochem. *30:* 571–578 (1972).

39 BIENZ, K.; EGGER, D., and WOLF, D. A.: Virus replication, cytopathology, and lysosomal enzyme response of mitotic and interphase Hep-2 cells infected with poliovirus. J. Virol. *11:* 565–574 (1973).

40 BIKEL, I. and KNIGHT, C. A.: Differential action of Aspergillus glycosidases on the hemagglutinating and neuraminidase activities of influenza and Newcastle disease virus. Virology *49:* 326–332 (1972).

41 BISHOP, D. H. L.; OBIJESKI, J. F., and SIMPSON, R. W.: Transcription of the influenza ribonucleic acid genome by a virion polymerase. J. Virol. *8:* 66–73 (1971).

42 BISHOP, D. H. L. and ROY, P.: Kinetics of RNA synthesis by vesicular stomatitis virus particles. J. molec. Biol. *57:* 513–527 (1971).

43 BISHOP, D. H. L.; ROY, P.; BEAN, W. J., and SIMPSON, R. W.: Transcription of the influenza ribonucleic acid genome by a virion polymerase. III. Completeness of the transcription process. J. Virol. *10:* 689–697 (1972).

44 BLASKOVIC, P.; RHODES, A. J.; DOANE, F. W., and LABZOFFSKY, N. A.: Infection of chick embryo tracheal organ cultures with influenza A2 (Hong Kong) virus. Arch. ges. Virusforsch. *38:* 250–266 (1972).

45 BLUMENFELD, O. O.; GALLOP, P. M.; HOWE, C., and LEE, L. T.: Erythrocyte membrane proteins. Their study using aqueous pyridine solutions. Biochim. biophys. Acta *211:* 109–123 (1970).

46 BOCHE, R. D. and QUILLIGAN, J. J.: Adsorption to glass and specific antibody inhibition of iodine$^{125}$ labeled influenza virus. J. Immunol. *97:* 942–950 (1966).

47 BODO, G.; SCHEIRER, W.; SUH, M.; SCHULTZE, B.; HORAK, I., and JUNGWIRTH, C.: Protein synthesis in pox-infected cells treated with interferon. Virology *50:* 140–147 (1972).

48 BOLOGNESI, D. P. and BAUER, H.: Polypeptides of avian RNA tumor viruses. I. Isolation and physical and chemical analysis. Virology *42:* 1097–1112 (1970).

49 BOLOGNESI, D. P.; BAUER, H.; GELDERBLOM, H., and HUPER, G.: Polypeptides of avian RNA tumor viruses. IV. Components of the viral envelope. Virology *47:* 551–566 (1972).

50 BOLOGNESI, D. P.; GELDERBLOM, H.; BAUER, H.; MOLLING, K., and HUPER, G.: Polypeptides of avian RNA tumor viruses. V. Analysis of the virus core. Virology *47:* 567–578 (1972).

51 Boone, C. W.; Irving, D. N., and Rubinstein, S.: Quantitative studies on the binding of antibody to the surface of HeLa cells. J. Immunol. *106:* 879–887 (1971).

52 Borsa, J.; Copps, T. P.; Sargent, M. D.; Long, D. G., and Chapman, J. D.: New intermediate subviral particles in the in vitro uncoating of reovirus virions by chymotrypsin. J. Virol. *11:* 552–564 (1973).

53 Borsa, J. and Graham, A. F.: Reovirus: RNA polymerase activity in purified virions. Biochem. biophys. Res. Commun. *33:* 895–901 (1968).

54 Borsa, J.; Sargent, M. D.; Long, D. G., and Chapman, J. D.: Extraordinary effects of specific monovalent cations on activation of reovirus transcriptase by chymotrypsin in vitro. J. Virol. *11:* 207–217 (1973).

55 Bose, H. R. and Sagik, B. P.: The virus envelope in cell attachment. J. gen. Virol. *9:* 159–161 (1970).

56 Bosmann, H. B.: Cellular membranes. Membrane marker enzyme activities in synchronized mouse leukemic cells L5178Y. Biochim. biophys. Acta *203:* 256–260 (1970).

57 Boulanger, P. A.; Breynaert, M. D., and Biserte, G.: Lysosomes and the problem of adenovirus uncoating. Exp. molec. Path. *12:* 235–242 (1970).

58 Boulanger, P. A.; Houdref, N.; Scharfman, A., and Lemay, P.: The role of surface sialic acid in adenovirus-cell adsorption. J. gen. Virol. *16:* 429–434 (1972).

59 Bourgaux, P.: The fate of polyoma virus in hamster, mouse, and human cells. Virology *23:* 46–55 (1964).

60 Boyd, A. L.; Butel, J. S.; Wong, C., and Melnick, J. L.: Adsorption of poliovirus by hamster cells after exposure to simian cell DNA. Intervirology (in press 1974).

61 Brady, R. O. and Mora, P. T.: Alteration in ganglioside pattern and synthesis in $SV_{40}$ and polyoma virus transformed mouse cell lines. Biochim. biophys. Acta *218:* 308–319 (1970).

62 Brand, K. G. and Syverton, J. T.: Immunology of cultivated mammalian cells. I. Species specificity determined by hemagglutination. J. nat. Cancer Inst. *24:* 1007–1019 (1960).

63 Breindl, M.: The structure of heated poliovirus particles. J. gen. Virol. *11:* 147–156 (1971).

64 Breindl, M.: VP4, the D-reactive part of poliovirus. Virology *46:* 962–964 (1971).

65 Breindl, M. and Koch, G.: Competence of suspended HeLa cells for infection by inactivated poliovirus particles and by isolated viral RNA. Virology *48:* 136–144 (1972).

65a Bretscher, M. S.: Major human erythrocyte glycoprotein spans the cell membrane. Nature new Biol. *231:* 229–232 (1971).

66 Brishammar, S. and Philipson, L.: Purification and properties of a non-specific hemagglutination inhibitor from human serum against picornavirus. Biochim. biophys. Acta *127:* 140–150 (1966).

66a Brown, D. T. and Burlingham, B. T.: Penetration of host cell membranes by adenovirus 2. J. Virol. *12:* 386–396 (1973).

67 Brown, F.; Cartwright, B., and Stewart, D. L.: Mechanism of infection of pig-kidney cells by foot-and-mouth disease virus. Biochim. biophys. Acta *47:* 172–177 (1961).

68 Brown, F.; Cartwright, B., and Stewart, D. L.: Further studies on the infection of pig-kidney cells by foot-and-mouth disease virus. Biochim. biophys. Acta *55:* 768–774 (1962).

69 Brubaker, M. M.; West, B., and Ellis, R. J.: Human blood group influence on reovirus hemagglutination titers. Proc. Soc. exp. Biol. Med. *115:* 1118–1120 (1964).

70 Buck, C. A.; Glick, M. C., and Warren, L.: Effect of growth on the glycoproteins from the surface of control and Rous sarcoma virus transformed hamster cells. Biochemistry *10:* 2176–2180 (1971).

71 Buck, C. A.; Glick, M. C., and Warren, L.: Glycopeptides from the surface of control and virus-transformed cells. Science *172:* 169–171 (1971).

71a Buck, C. A.; Granger, G. A.; Taylor, M. W., and Holland, J. J.: Efficient, inefficient, and abortive infection of different mammalian cells by small RNA viruses. Virology *33:* 36–46 (1967).

72 Buckland, F. E.: Inactivation of virus hemagglutinins by para-chloromercuribenzoic acid. Nature, Lond. *188:* 768 (1960).

73 Burge, B. W. and Huang, A. S.: Comparison of membrane protein glycopeptides of Sindbis virus and vesicular stomatitis virus. J. Virol. *6:* 176–182 (1970).

74 Burness, A. T. H.; Pardoe, I. V., and Fox, S. M.: Evidence for the lack of glycoprotein in the encephalomyocarditis virus particle. J. gen. Virol. *18:* 33–49 (1973).

75 Burroughs, J. N.; Rowlands, D. J.; Sanger, D. V.; Talbot, P., and Brown, F.: Further evidence for multiple proteins in the foot-and-mouth disease virus particle. J. gen. Virol. *13:* 73–84 (1971).

76 Caliguiri, L. A. and Tamm, I.: The role of cytoplasmic membranes in poliovirus biosynthesis. Virology *42:* 100–111 (1970).

77 Came, P. E. and Crowell, R. L.: Studies of resistance of fetal mouse tissue in culture to coxsackie group A viruses. Virology *23:* 542–552 (1964).

78 Capaldi, R. A. and Vanderkooi, G.: The low polarity of many membrane proteins. Proc. nat. Acad. Sci., Wash. *69:* 930–932 (1972).

78a Carey, F. J.; Kuhn, N. O., and Harford, C. G.: Effects of anticellular serum on phagocytosis and the uptake of tritiated thymidine and uridine by HeLa cells. J. exp. Med. *121:* 991–999 (1965).

79 Carter, S. B.: Effects of cytochalasins on mammalian cells. Nature, Lond. *213:* 261–264 (1967).

80 Cartwright, B.; Smale, C. J., and Brown, F.: Structural and biological relations in vesicular stomatitis virus; in Barry and Mahy The biology of large RNA viruses, pp. 115–132 (Academic Press, New York 1970).

81 Cartwright, B.; Smale, C. J., and Brown, F.: Dissection of vesicular stomatitis virus into the infective ribonucleoprotein and immunizing components. J. gen. Virol. *7:* 19–32 (1970).

82 Cartwright, S. F. and Thorne, H. V.: Use of two properties of the virus of foot-and-mouth disease in studies of its multiplication. Nature, Lond. *182:* 717–718 (1958).

83 Cartwright, S. F. and Thorne, H. V.: Some applications of detergents to the study of the virus of foot-and-mouth disease. J. gen. Microbiol. *20:* 61–77 (1959).

84 Carubelli, R.; Bhavanandan, V. P., and Gottschalk, A.: Studies on glycoproteins. XI. The O-glycosidic linkage of N-acetylgalactosamine to seryl and threonyl residues in ovine submaxillary gland glycoprotein. Biochim. biophys. Acta *101:* 67–82 (1965).

85 Caspar, D. L. D.: Assembly and stability of the tobacco mosaic virus particle. Adv. Protein Chem. *18:* 37–121 (1963).

86 Caspar, D. L. D.: Design principles in organized biological structures. Ciba Found. Symp. 1965; in Wolstenholme and O'Connor Principles of biomolecular organization, pp. 7–39 (Churchill, London 1966).

87 Caspar, D. L. D. and Klug, A.: Physical principles in the construction of regular viruses. Cold Spr. Harb. Symp. quant. Biol. *27:* 1–24 (1962).

88 Casto, B. C. and Hammon, W. M.: Correlation between propagation of a bovine enterovirus and development of receptor-like material in tissue mince. Proc. Soc. exp. Biol. Med. *132:* 154–160 (1969).

89 Ceglowski, W. S.: Antibody response to the noninfectious 7 mμ component of the virus of foot-and-mouth disease. Virology *25:* 328–330 (1965).

90 Chan, V. F. and Black, F. L.: Uncoating of poliovirus by isolated plasma membranes. J. Virol. *5:* 309–312 (1970).

91 Chang, C. T. and Zweerink, H. J.: Fate of parental reovirus in infected cell. Virology *46:* 544–555 (1971).

92 Chardonnet, Y. and Dales, S.: Early events in the interaction of adenoviruses with HeLa cells. I. Penetration of type 5 and intracellular release of the DNA genome. Virology *40:* 462–477 (1970).

93 Chardonnet, Y. and Dales, S.: Early events in the interaction of adenoviruses with HeLa cells. II. Comparative observations on the penetration of type 1, 5, 7 and 12. Virology *40:* 478–485 (1970).

94 Chardonnet, Y. and Dales, S.: Early events in the interaction of adenoviruses with HeLa cells. III. Relationship between an ATPase activity in nuclear envelopes and transfer of core material. A hypothesis. Virology *48:* 342–359 (1972).

95 Chen, C.; Compans, R. W., and Choppin, P. W.: Parainfluenza virus surface projections. Glycoproteins with haemagglutinin and neuraminidase activities. J. gen. Virol. *11:* 53–58 (1971).

96 Choppin, P. W. and Tamm, I.: Studies of two kinds of virus particles which comprise influenza A2 virus strains. I. Characterization of stable homogeneous substrains in reactions with specific antibody, mucoprotein inhibitors, and erythrocytes. J. exp. Med. *112:* 895–920 (1960).

97 Choppin, P. W. and Tamm, I.: Studies of two kinds of virus particles which comprise influenza A2 virus strains. II. Reactivity with virus inhibitors in normal sera. J. exp. Med. *112:* 921–944 (1960).

98 Chow, N. L. and Simpson, R. W.: RNA-dependent RNA polymerase activity associated with virions and subviral particles of myxoviruses. Proc. nat. Acad. Sci., Wash. *68:* 752–756 (1971).

99 Chu, L. W. and Morgan, H. R.: Studies of the hemolysis of red blood cells by mumps virus. II. The relationships of hemagglutination, virus elution and hemolysis. J. exp. Med. *91:* 403–416 (1950).

100 Čiampor, F. and Križanová, O.: Interaction of plasma membranes with influenza virus. III. Electron microscopic study of interactions between influenza virus and isolated plasma membranes. Acta virol., Praha *15:* 361–366 (1971).

101 Clamp, J. R. and Hough, L.: The periodate oxidation of amino acids with reference to studies on glycoproteins. Biochem. J. *94:* 17–24 (1965).

102 Clarke, A. E. and Denborough, M. A.: The interaction of concanavalin A with blood-group-substance glycoproteins from human secretions. Biochem. J. *121:* 811–816 (1971).

103 Codington, J. F.; Sanford, B. H., and Jeanloz, R. W.: Glycoprotein coat of the $TA_3$ cell. I. Removal of carbohydrate and protein material from viable cells. J. nat. Cancer Inst. *45:* 637–647 (1970).

104 Coffin, J. M. and Temin, H. M.: Comparison of Rous sarcoma virus specific deoxyribonucleic acid polymerases in virions of Rous sarcoma virus and in Rous sarcoma virus-infected chicken cells. J. Virol. *7:* 625–634 (1971).

105 Cohn, Z. A.: The regulation of pinocytosis in mouse macrophages. I. Metabolic requirements as defined by the use of inhibitors. J. exp. Med. *124:* 557–571 (1966).

106 Cole, C. N. and Baltimore, D.: Defective interfering particles of poliovirus. II. Nature of the defect. J. molec. Biol. *76:* 325–343 (1973).

107 Cole, C. N.; Smoler, D.; Wimmer, E., and Baltimore, D.: Defective interfering particles of poliovirus. I. Isolation and physical properties. J. Virol. *7:* 478–485 (1971).

108 Colter, J. S. and Campbell, J. B.: The effect of polyanions and polycations on Mengo virus-L cell interaction. Ann. N. Y. Acad. Sci. *130:* 383–389 (1965).

109 Compans, R. W.: Influenza virus proteins. II. Association with components of the cytoplasm. Virology *51:* 56–70 (1973).

110 Compans, R. W. and Choppin, P. W.: Isolation and properties of the helical nucleocapsid of the parainfluenza virus SV5. Proc. nat. Acad. Sci., Wash. *57:* 949–956 (1967).

111 Compans, R. W. and Choppin, P. W.: The structure and assembly of influenza and parainfluenza viruses; in Maramorosch and Kurstak Comparative virology, pp. 407–432 (Academic Press, New York 1971).

112 Compans, R. W. and Dimmock, N. J.: An electron microscopic study of single-cycle infection of chick embryo fibroblasts by influenza virus. Virology *39:* 499–515 (1970).

113 Compans, R. W.; Klenk, H. P.; Caliguiri, L. A., and Choppin, P. W.: Influenza virus proteins. I. Analysis of polypeptides of the virion and identification of spike glycoproteins. Virology *42:* 880–889 (1970).

114 Cook, G. M. W.; Heard, D. H., and Seaman, G. A. F.: Sialic acids and the electrokinetic charge of the human erythrocyte. Nature, Lond. *191:* 44–47 (1961).

115 Cooper, P. D.: Synchrony and the elimination of chance delays in the growth of poliovirus. J. gen. Microbiol. *37:* 259–266 (1964).

116 Cooper, P. D. and Bellett, A. J. D.: A transmissible interfering component of vesicular stomatitis virus preparations. J. gen. Microbiol. *21:* 485–497 (1959).

117 Cordell-Stewart, B. and Taylor, M. W.: Effect of viral double-stranded RNA on protein synthesis in intact cells. J. Virol. *11:* 232–237 (1973).

118 Cowan, K. M. and Graves, J. H.: A third antigenic component associated with foot-and-mouth disease infection. Virology *30:* 528–540 (1966).

119 Crawford, L. V.: The adsorption of polyoma virus. Virology *18:* 177–181 (1962).

120 Crick, F. H. C. and Watson, J. D.: Structure of small viruses. Nature, Lond. *177:* 473–475 (1956).

121 Crittenden, L. B.; Briles, W. E., and Stone, H. A.: Susceptibility to an avian leukosis-sarcoma virus. Close association with an erythrocyte isoantigen. Science *169:* 1324–1325 (1970).

122 Crowell, R. L.: Specific viral interference in HeLa cell cultures chronically infected with coxsackie B5 virus. J. Bact. *86:* 515–526 (1963).

123 Crowell, R. L.: Specific cell-surface alteration by enteroviruses as reflected by viral-attachment interference. J. Bact. *91:* 198–204 (1966).

124 Crowell, R. L.: Determination of the relative receptor concentration of HeLa cells for different enteroviruses by attachment kinetics. Bact. Proc. *1967:* 140 (v. 35).

125 Crowell, R. L.: Effect of pH on attachment of enteroviruses to HeLa cells. Bact. Proc. *1968:* 180 (v. 212).

126 Crowell, R. L.; Landau, B. J., and Philipson, L.: The early interaction of coxsackievirus B3 with HeLa cells. Proc. Soc. exp. Biol. Med. *137:* 1082–1088 (1971).

127 Crowell, R. L. and Philipson, L.: Specific alterations of coxsackievirus B3 eluted from HeLa cells. J. Virol. *8:* 509–515 (1971).

128 Crowell, R. L. and Syverton, J. T.: The mammalian cell-virus relationship. VI. Sustained infection of HeLa cells by coxsackie B3 virus and effect on superinfection. J. exp. Med. *113:* 419–435 (1961).

129 Crowther, R. A.; Amos, L. A.; Finch, J. T.; de Rosier, D. J., and Klug, A.: Three dimensional reconstructions of spherical viruses by Fourier synthesis from electron micrographs. Nature, Lond. *226:* 421–425 (1970).

130 Cuatrecasas, P.: Insulin-receptor interactions in adipose tissue cells. Direct measurement and properties. Proc. nat. Acad. Sci., Wash. *68:* 1264–1268 (1971).

131 Cuatrecasas, P.: Isolation of the insulin receptor of liver and fat-cell membranes. Proc. nat. Acad. Sci., Wash. *69:* 318–322 (1972).

132 Cuatrecasas, P.: Affinity chromatography and purification of the insulin receptor of liver cell membranes. Proc. nat. Acad. Sci., Wash. *69:* 1277–1281 (1972).

133 Cumar, F. A.; Brady, R. O.; Kolodny, E. H.; McFarland, V. W., and Mora, P. T.: Enzymatic block in the synthesis of gangliosides in DNA virus-transformed tumorigenic mouse cell lines. Proc. nat. Acad. Sci., Wash. *67:* 757–764 (1970).

134 Dahl, R. and Kates, J. R.: Intracellular structures containing vaccinia DNA. Isolation and characterization. Virology *42:* 453–462 (1970).

135 Dahl, R. and Kates, J. R.: Synthesis of vaccinia virus 'early' and 'late' messenger RNA *in vitro* with nucleoprotein structures isolated from infected cells. Virology *42:* 463–472 (1970).

136 Dales, S.: An electron microscope study of the early association between two mammalian viruses and their hosts. J. Cell Biol. *13:* 303–322 (1962).

137 Dales, S.: The uptake and development of vaccinia virus in strain L cells followed with labeled viral deoxyribonucleic acid. J. Cell Biol. *18:* 51–72 (1963).

138 Dales, S.: Penetration of animal viruses into cells; in Melnick Progr. med. Virol., vol. 7, pp. 1–43 (Karger, Basel 1965).

139 Dales, S.: The early events in cell-animal virus interactions. Bact. Rev. *37:* 103–135 (1973).

140 Dales, S.; Hanafusa, H., and Huima, T.: Penetration and intracellular release of the genomes of avian RNA tumor viruses. Virology *50:* 440–458 (1972).

141 Dales, S. and Kajioka, R.: The cycle of multiplication of vaccinia virus in Earle's strain L cells. I. Uptake and penetration. Virology *24:* 278–294 (1964).

142 Dales, S. and Silverberg, H.: Viropexis of herpes simplex virus by HeLa cells. Virology *37:* 475–480 (1969).

143 Dalgarno, L.; Cox, R. A., and Martin, E. M.: Polyribosomes in normal Krebs 2 ascites tumor cells and in cells infected with encephalomyocarditis virus. Biochim. biophys. Acta *138:* 316–328 (1967).

144 Danon, D.; Howe, C., and Lee, L. T.: Interaction of polylysine with soluble components of human erythrocyte membranes. Biochim. biophys. Acta *101:* 201–213 (1965).

145 Danziger, R. E. and Paranchych, W.: Stages in phage R17 infection. II. Ionic requirements for phage R17 attachment to F-pili. Virology *40:* 547–553 (1970).

146 Danziger, R. E. and Paranchych, W.: Stages in phage R17 infection. III. Energy requirements for the F-pili mediated eclipse of viral infectivity. Virology *40:* 554–564 (1970).

147 Darnbrough, C.; Hunt, T., and Jackson, R. J.: A complex between met-t-RNA and native 40s subunits in reticulocyte lysates and its disappearance during incubation with double-stranded RNA. Biochem. biophys. Res. Commun. *48:* 1556–1564 (1972).

148 Darnell, J. E.: Adsorption and maturation of poliovirus in singly and multiply infected HeLa cells. J. exp. Med. *107:* 633–641 (1958).

149 Darnell, J. E.; Girard, M.; Baltimore, D.; Summers, D. F., and Maizel, J. V.: The synthesis and translation of poliovirus RNA; in Colter and Paranchych The molecular biology of viruses, pp. 375–402 (Academic Press, New York 1967).

150 Darnell, J. E. and Sawyer, T. K.: The basis for variation in susceptibility to poliovirus in HeLa Cells. Virology *11:* 665–675 (1960).

151 Da Silva, P. P.: Translational mobility of the membrane intercalated particles of human erythrocyte ghosts (pH-dependent, reversible aggregation). J. Cell Biol. *53:* 777–787 (1972).

152 David-Ferreira, J. F. and Manaker, R. A.: An electron microscope study of the development of a mouse hepatitis virus in tissue culture cells. J. Cell Biol. *24:* 57–78 (1965).

153 Davis, A. T.; Estensen, R. D., and Quie, P. G.: Cytochalasin B. III. Inhibition of human polymorphonuclear leukocyte phagocytosis. Proc. Soc. exp. Biol. Med. *137:* 161–164 (1971).

154 Davies, W. L.; Grunert, R. R.; Haff, R. F.; McGahen, J. W.; Neumayer, E. M.; Paulshock, M.; Watts, J. C.; Wood, T. R.; Herrmann, E. C., and Hoffmann, C. E.: Antiviral activity of 1-adamantanamine (amantadine). Science *144:* 862–863 (1964).
155 Vide Verlinde and de Baan [693].
156 Defendi, V. and Jensen, F.: Oncogenicity by DNA tumor viruses. Enhancement after ultraviolet and cobalt-60 radiations. Science *157:* 703–705 (1967).
157 Deitch, A. D. and Godman, G. C.: Enhancement of viral infectivity by cytochalasin D. Abstracts of papers 12th Ann. Meet. American Society for Cell Biology. J. Cell Biol. *55:* 58a (1972).
158 Deitch, A. D.; Godman, G. C., and Tanenbaum, S. W.: Cytochalasin D enhances infectivity of poliovirus. Proc. Soc. exp. Biol. Med. *144:* 60–64 (1973).
159 De Petris, S. and Raft, M. C.: Normal distribution, patching and capping of lymphocyte surface immunoglobulin studied by electron microscopy. Nature new Biol. *241:* 257–259 (1973).
160 Diderholm, H. and Dinter, Z.: A line of polyoma-transformed BHK21 cells insusceptible to foot-and-mouth disease virus. Virology *26:* 369–372 (1965).
161 Dourmashkin, R. R. and Tyrrell, D. A. J.: Attachment of two myxoviruses to ciliated epithelial cells. J. gen. Virol. *9:* 77–88 (1970).
162 Dowell, C. E.: Cold-sensitive mutants of bacteriophage ØX-174. I. A mutant blocked in the eclipse function at low temperature. Proc. nat. Acad. Sci., Wash. *58:* 958–961 (1967).
163 Drzeniek, R.: Differences in splitting capacity of virus and V. cholerae neuraminidases on sialic acid type substrates. Biochem. biophys. Res. Commun. *26:* 631–638 (1967).
164 Drzeniek, R.: Wechselwirkungen zwischen Myxoviren und Zelloberflächen. III. Neuraminidase der Myxoviren. Z. med. Microbiol. Immunol. *156:* 1–38 (1970).
165 Drzeniek, R.; Frank, H., and Rott, R.: Electron microscopy of purified influenza virus neuraminidase. Virology *36:* 703–707 (1968).
166 Duc-Nguyen, H.: Enhancing effect of diethylaminoethyl-dextran on the focus-forming titer of a murine sarcoma virus (Harvey strain). J. Virol. *2:* 643–644 (1968).
167 Duc-Nguyen, H. and Henle, W.: Replication of mumps virus in human leukocyte cultures. J. Bact. *92:* 258–265 (1966).
168 Dudley, M. A.; Henkens, R. W., and Rowlands, D. T.: Kinetics of neutralization of bacteriophage $f_2$ by rabbit $\gamma$G-antibodies. Proc. nat. Acad. Sci., Wash. *65:* 88–95 (1970).
169 Duesberg, P. H.; Martin, G. S., and Vogt, P. K.: Glycoprotein components of avian and murine RNA tumor viruses. Virology *41:* 631–646 (1970).
170 Dulbecco, R.: Cell transformation by viruses. Two minute viruses are powerful tools for analyzing the mechanism of cancer. Science *166:* 962–968 (1969).
171 Dulbecco, R. and Eckhart, W.: Temperature-dependent properties of cells transformed by a thermosensitive mutant of polyoma virus. Proc. nat. Acad. Sci., Wash. *67:* 1775–1781 (1970).
172 Dulbecco, R.; Hariwell, L. H., and Vogt, M.: Induction of cellular DNA synthesis by polyoma virus. Proc. nat. Acad. Sci., Wash. *53:* 403–410 (1965).

173 DULBECCO, R. and JOHNSON, T.: Interferon-sensitivity of the enhanced incorporation of thymidine into cellular DNA induced by polyoma virus. Virology *42:* 368–374 (1970).

174 DULBECCO, R. and STOKER, M. G. P.: Conditions determining initiation of DNA synthesis in 3T3 cells. Proc. nat. Acad. Sci., Wash. *66:* 204–210 (1970).

175 DUNKER, K. A. and RUECKERT, R. R.: Fragments generated by pH dissociation of ME-virus and their relation to the structure of the virion. J. molec. Biol. *58:* 217–235 (1971).

176 DUNNEBACKE, T. H.; LEVINTHAL, J. D., and WILLIAMS, R. C.: Entry and release of poliovirus as observed by electron microscopy of cultured cells. J. Virol. *4:* 505–513 (1969).

177 DURAND, D. P.; BROTHERTON, T.; CHALGREN, S., and COLLARD, W.: Uncoating of myxoviruses; in BARRY and MAHY The biology of large RNA viruses, pp. 197–206 (Academic Press, New York 1970).

178 DZIERZKOWA-BORODEJ, W.; LISOWSKA, E., and SEYFRIEDOWA, H.: The activity of glycoproteins from erythrocytes and protein fractions of human colostrum towards anti-I antibodies. Life Sci. *9:* 111–120 (1970).

179 EASTERBROOK, K. B.: Controlled degradation of vaccinia virions *in vitro:* an electron microscopic study. J. Ultrastruct. Res. *14:* 484–496 (1966).

180 EDELMAN, R. and WHEELOCK, E. F.: Specific role of each human leukocyte type in viral infections. II. Phytohemagglutinin-treated lymphocytes as host cells for vesicular stomatitis virus replication in vitro. J. Virol. *2:* 440–448 (1968).

181 EGGERS, H. J.: Inhibition of early stages of virus-cell interactions. Ann. N. Y. Acad. Sci. *173:* 417–419 (1970).

182 EHRENFELD, E. and HUNT, T.: Double-stranded poliovirus RNA inhibits initiation of protein synthesis by reticulocyte lysates. Proc. nat. Acad. Sci., Wash. *68:* 1075–1078 (1971).

183 ELSBACH, P.; HOLMES, K. V., and CHOPPIN, P. W.: Metabolism of lecithin and virus-induced cell fusion. Proc. Soc. exp. Biol. Med. *130:* 903–908 (1969).

184 ENDERS, F.; HOLLOWAY, A., and GROGAN, E. A.: Replication of poliovirus 1 in chick embryo and hamster cells exposed to Sendai virus. Proc. nat. Acad. Sci., Wash. *57:* 637–644 (1967).

185 EPSTEIN, M. A. and HOLT, S. J.: Electron microscopic observations on the surface adenosine triphosphatase-like enzymes of HeLa cells infected with herpes virus. J. Cell Biol. *19:* 337–347 (1963).

186 EPSTEIN, M. A.; HUMMELER, K., and BERKALOFF, A.: The entry and distribution of herpes virus and colloidal gold in HeLa cells after contact in suspension. J. exp. Med. *119:* 291–312 (1964).

187 EREMENKO, T.; BENEDETTO, A., and VOLPE, P.: Virus infection as a function of the host cell life cycle. Replication of poliovirus RNA. J. gen. Virol. *16:* 61–68 (1972).

188 EREMENKO, T.; BENEDETTO, A., and VOLPE, P.: Poliovirus replication during HeLa cell life cycle. Nature new Biol. *237:* 114–116 (1972).

189 ESTEBAN, M. and METZ, D. H.: Early virus protein synthesis in vaccinia virus-infected cells. J. gen. Virol. *19:* 201–216 (1973).

190 Estensen, R. D. and Plagemann, P. G. W.: Cytochalasin B. Inhibition of glucose and glucosamine transport. Proc. nat. Acad. Sci., Wash. *69:* 1430–1434 (1972).

191 Eustatia, J. M. and Van der Veen, J.: Viral replication in cultures of phytohemagglutinin-treated mouse lymphocytes. Proc. Soc. exp. Biol. Med. *137:* 424–428 (1971).

192 Evans, M. J. and Kingsbury, D. W.: Separation of Newcastle disease virus proteins by polyacrylamide gel electrophoresis. Virology *37:* 597–604 (1969).

193 Everitt, E.; Sundquist, B.; Pettersson, U., and Philipson, L.: Structural proteins of adenoviruses. X. Isolation and topography of low molecular weight antigens from the virion of adenovirus type 2. Virology *52:* 130–147 (1973).

194 Farnham, A. E. and Newton, A. A.: The effect of some environmental factors on herpes virus grown in HeLa cells. Virology *7:* 449–461 (1959).

195 Fazekas de St. Groth, S.: Viropexis, the mechanism of influenza virus infection. Nature, Lond. *162:* 294–295 (1948).

196 Fazekas de St. Groth, S. and Gottschalk, A.: Studies on glycoproteins. X. Equilibrium measurements on influenza virus-glycoprotein systems. Biochim. biophys. Acta *78:* 248–257 (1963).

197 Fazekas de St. Groth, S. and Webster, R. G.: The neutralization of animal viruses. III. Equilibrium conditions in the influenza virus-antibody system. J. Immunol. *90:* 140–150 (1963).

198 Fazekas de St. Groth, S. and Webster, R. G.: The neutralization of animal viruses. IV. Parameters of the influenza virus-antibody system. J. Immunol. *90:* 151–164 (1963).

199 Fenwick, M. L.: The influence of poliovirus infection on RNA synthesis in mammalian cells. Virology *19:* 241–249 (1963).

200 Fenwick, M. L. and Cooper, P. D.: Early interactions between poliovirus and ERK cells. Some observations on the nature and significance of the rejected particles. Virology *18:* 212–223 (1962).

201 Finch, J. T. and Gibbs, A. J.: Observations on the structure of the nucleocapsids of some paramyxoviruses. J. gen. Virol. *6:* 141–150 (1970).

202 Fogh, J. and Fogh, H.: Chromosome changes in PPLO-infected FL human amnion cells. Proc. Soc. exp. Biol. Med. *119:* 223–238 (1965).

203 Fong, C. K. Y.; Bensch, K. G., and Hsiung, G. D.: Productive and abortive infections of simian and nonsimian cells with a simian adenovirus, SV15. I. Microscopic observations. Virology *35:* 297–310 (1968).

204 Fong, C. K. Y.; Hsiung, G. D., and Bensch, K. G.: Productive and abortive infections of simian and nonsimian cells with a simian adenovirus, SV15. II. Viral biosynthesis and effects of chemical inhibitors. Virology *35:* 311–320 (1968).

205 Forsgren, M.: Relationship between poliovirus and echovirus 6 antigens. I. Immunization experiments in guinea pigs. Arch. ges. Virusforsch. *39:* 108–120 (1972).

206 Fox, T. O.; Sheppard, J. R., and Burger, M. M.: Cyclic membrane changes in animal cells: Transformed cells permanently display a surface architecture detected in normal cells only during mitosis. Proc. nat. Acad. Sci., Wash *68:* 244–247 (1971).

207 FRANKLIN, R. M. and BALTIMORE, D.: Patterns of macromolecular synthesis in normal and virus-infected mammalian cells. Cold Spr. Harb. Symp. quant. Biol. *27:* 175–195 (1962).

208 FREARSON, P. M. and CRAWFORD, L. V.: Polyoma virus basic proteins. J. gen. Virol. *14:* 141–155 (1972).

209 FRIED, M. and PITTS, J. D.: Replication of polyoma virus DNA. I. A resting cell system for biochemical studies on polyoma virus. Virology *34:* 761–770 (1968).

210 FRIEDMAN, R. M.: Structural and nonstructural proteins of an arbovirus. J. Virol. *2:* 1076–1080 (1968).

211 FRISCH-NIGGEMEYER, W.: Polyphosphoinositides as receptor substances for certain groups of arboviruses. Acta virol., Praha *15:* 119–125 (1971).

211a FRISCH-NIGGEMEYER, W. and LEITNER, H.: Competitive inhibition of hemagglutination of tick-borne encephalitis virus by receptor-analogue substances. Acta virol., Praha *17:* 264 (1973).

212 FRYE, L. D. and EDIDIN, M.: The rapid intermixing of cell surface antigens after formation of mouse-human heterokaryons. J. Cell Sci. *7:* 319–335 (1970).

213 FURLONG, D.; SWIFT, H., and ROIZMAN, B.: Arrangement of herpesvirus deoxyribonucleic acid in the core. J. Virol. *10:* 1071–1074 (1972).

214 GAGNONI, G.; CANTAGALLI, P.; GRASSO, A., and DIANZANI, F.: Effect of reagents affecting specific groups of proteins on the haemagglutinating activity of rubella virus. Acta virol., Praha *13:* 235–238 (1969).

215 GARD, S. and HELLER, L.: Hemagglutination by Col-MM-virus. Proc. Soc. exp. Biol. Med. *76:* 68–73 (1951).

216 GELB, L. D. and LERNER, A. M.: Reovirus hemagglutination inhibition by N-acetyl-D-glucosamine. Science *147:* 404–405 (1965).

217 GERSHON, D.; HAUSEN, P.; SACHS, L., and WINOCOUR, E.: On the mechanism of polyoma virus-induced synthesis of cellular DNA. Proc. nat. Acad. Sci., Wash. *54:* 1584–1592 (1965).

218 GERSHON, D.; SACHS, L., and WINOCOUR, E.: The induction of cellular DNA synthesis by simian virus 40 in contact-inhibited and in X-irradiated cells. Proc. nat. Acad. Sci., Wash. *56:* 918–925 (1966).

219 GHENDON, Y. Z. and YACOBSON, E. A.: Antigenic specificity of poliovirus-related particles. J. Virol. *8:* 589–590 (1971).

220 GIBSON, W. and ROIZMAN, B.: Proteins specified by herpes simplex virus. VIII. Characterization and composition of multiple capsid forms of subtypes 1 and 2. J. Virol. *10:* 1044–1052 (1972).

221 GINSBERG, H. S.; BELLO, L. J., and LEVINE, A. J.: Control of biosynthesis of host macromolecules in cells infected with adenoviruses; in COLTER and PARANCHYCH Molecular biology of viruses, pp. 547–572 (Academic Press, New York 1967).

222 GINSBERG, H. S.; PEREIRA, H. G.; VALENTINE, R. C., and WILCOX, W. C.: A proposed terminology for the adenovirus antigens and virion morphological subunits. Virology *28:* 782–783 (1966).

223 GINSBERG, V. and KOBATA, A.: Structure and function of surface components of mammalian cells; in ROTHFIELD Structure and function of biological membranes, pp. 439–459 (Academic Press, New York 1971).

223a GINSBERG, V.; KOBATA, A.; HICKEY, C., and SAWICKA, T.: Biochemical basis for blood types in man; in JAMIESON and GREENWALT Glycoproteins of blood cells and plasma, pp. 114–126 (Lippincott, Philadelphia 1971).

224 GOLDBERG, R. J. and CROWELL, R. L.: Susceptibility of differentiating muscle cells of the fetal mouse in culture to coxsackievirus A13. J. Virol. *7:* 759–769 (1971).

225 GOLDBERG, R. J.; GRAVELL, M., and CROWELL, R. L.: Evidence for the susceptibility of primary fetal mouse cells in culture to coxsackievirus A13. Proc. Soc. exp. Biol. Med. *132:* 743–748 (1969).

226 GOLDFIELD, M.; SRIHONGSE, S., and FOX, J. P.: Hemagglutinins associated with certain human enteric viruses. Proc. Soc. exp. Biol. Med. *96:* 788–791 (1957).

227 GOLDSTEIN, E. A. and PONS, M. W.: The effects of polyvinylsulfate on the ribonucleoprotein of influenza virus. Virology *41:* 382–384 (1970).

228 GOMATOS, P. J. and TAMM, I.: Reactive sites of reovirus type 3 and their interaction with receptor substances. Virology *17:* 455–461 (1962).

229 GOODHEART, C. R.: Non-transfer of RNA from parent to progeny of encephalomyocarditis virus. J. molec. Biol. *23:* 183–190 (1967).

230 GORMAN, B.: Lipid inhibitors of arbovirus haemagglutination. J. gen. Virol. *6:* 305–313 (1970).

231 GOTTSCHALK, A.: Influenza virus haemagglutinin inhibitors; in GOTTSCHALK Glycoproteins, pp. 543–547 (Elsevier, Amsterdam 1966).

232 GRAHAM, E. R. B. and GOTTSCHALK, A.: Studies on mucoproteins. I. The structure of the prosthetic group of ovine submaxillary gland mucoprotein. Biochim. biophys. Acta *38:* 513–524 (1960).

233 GRANADOS, R. R.: Entry of an insect poxvirus by fusion of the virus envelope with the host cell membrane. Virology *52:* 305–309 (1972).

234 GRAVES, J. H.; COWAN, K. M., and TRAUTMAN, R.: Immunochemical studies of foot-and-mouth disease. II. Characterization of RNA-free viruslike particles. Virology *34:* 269–274 (1968).

235 GREEN, M.; PIÑA, M., and KIMES, R. C.: Biochemical studies on adenovirus multiplication. XII. Plaquing efficiencies of purified human adenoviruses. Virology *31:* 562–565 (1967).

236 GREEN, H.; WANG, R.; BASILICO, C.; POLLACK, R.; KUSANO, T., and SALES, J.: Mammalian somatic cell hybrids and their susceptibility to viral infection. Fed. Proc. *30:* 930–934 (1971).

237 GUIDOTTI, G.: Membrane proteins. Ann. Rev. Biochem. *41:* 731–752 (1972).

238 GYÖRGY, E. and KOCH, A.: Heart glycosides in poliovirus-host cell interaction. II. Effect of the time of addition on stimulatory or inhibitory action. Acta microbiol. hung. *16:* 197–202 (1969).

239 HABEL, K.; HORNIBROOK, J. W.; GREGG, N. C.; SILVERBERG, R. J., and TAKEMOTO, K. K.: The effect of anticellular sera on virus multiplication in tissue culture. Virology *5:* 7–29 (1958).

240 HAFF, R. F.; WOHLSN, B.; FORCE, E. E., and STEWART, R. C.: Growth characteristics of two rhinovirus strains in WI-26 and monkey kidney cells. J. Bact. *91:* 2339–2342 (1966).

241 HAHN, E. and FOGH, J.: Increased resistance of SV40 transformed human amnion cells to poliovirus infection. Arch. ges. Virusforsch. *29:* 343–360 (1970).

242 HAKOMORI, S.: Cell density-dependent changes of glycolipid concentration in fibroblasts, and loss of this response in virus-transformed cells. Proc. nat. Acad. Sci., Wash. *67:* 1741–1747 (1970).

243 HAKOMORI, S. I.; SAITO, T., and VOGT, P. K.: Transformation by Rous sarcoma virus. Effects on cellular glycolipids. Virology *44:* 609–621 (1971).

244 HALL, L. and RUECKERT, R.: Infection of mouse fibroblasts by cardioviruses. Premature uncoating and its prevention by elevated pH and magnesium chloride. Virology *43:* 152–165 (1971).

245 HALPEREN, S.; EGGERS, H. J., and TAMM, I.: Complete and coreless hemagglutinating particles produced in ECHO 12 virus-infected cells. Virology *23:* 81–89 (1964).

246 HALPEREN, S.; SCHIEBLE, J. H.; MORI, R.; DINKA, S., and ACKERMANN, W. W.: Some properties of an inhibitor of polyoma virus hemagglutination isolated from bovine serum. J. Immunol. *89:* 400–407 (1962).

247 HALPEREN, S.; STONE, H. O., and KORANT, B. D.: Isolation of glucosamine from the capsids of a picornavirus. J. gen. Virol. *20:* 267–276 (1973).

248 HANAFUSA, T. and HANAFUSA, H.: Interaction among avian tumor viruses giving enhanced infectivity. Proc. nat. Acad. Sci., Wash. *58:* 818–825 (1967).

249 HAND, R. and TAMM, I.: Reovirus. Effect of noninfective viral components on cellular deoxyribonucleic acid synthesis. J. Virol. *11:* 223–231 (1973).

250 HANIG, M.: Electrokinetic change in human erythrocytes during adsorption and elution of PR8 influenza virus. Proc. Soc. exp. Biol. Med. *68:* 385–392 (1948).

251 HARRAP, K. A.: Cell infection by a nuclear polyhedrosis virus. Virology *42:* 311–318 (1970).

252 HARRIS, J. I. and HINDLEY, J.: The protein subunit of turnip yellow mosaic virus. J. molec. Biol. *3:* 117–120 (1961).

253 HARRISON, S. C.; DAVID, A.; JUMBLATT, J., and DARNELL, J. E.: Lipid and protein organization in Sindbis virus. J. molec. Biol. *60:* 523–528 (1971).

254 HARTER, D. H. and CHOPPIN, P. W.: Adsorption of attenuated and neurovirulent poliovirus strains to central nervous system tissues of primates. J. Immunol. *95:* 730–736 (1965).

255 HARTLEY, J. W.; ROWE, W. P.; CHANOCK, R. M., and ANDREWS, B. E.: Studies of mouse polyoma virus infection. IV. Evidence for mucoprotein erythrocyte receptors in polyoma virus hemagglutination. J. exp. Med. *110:* 81–91 (1959).

256 HATANAKA, M.; KAKEFUDA, T.; GILDEN, R. V., and CALLAN, E. A. O.: Cytoplasmic DNA synthesis induced by RNA tumor viruses. Proc. nat. Acad. Sci., Wash. *68:* 1844–1847 (1971).

257 HAWKES, R. A. and LAFFERTY, K. J.: The enhancement of virus infectivity by antibody. Virology *33:* 250–261 (1967).

258 HAY, A. J.; SKEHEL, J. J., and BURKE, D. C.: Proteins synthesized in chick cells following infection with Semliki Forest virus. J. gen. Virol. *3:* 175–184 (1968).

259 HEATH, E. C.: Complex polysaccharides. Annu. Rev. Biochem. *40:* 29–56 (1971).

260 HEBERLING, R. L. and CHEEVER, F. S.: Simian enterovirus SV2 hemagglutination studies. II. Effect of various treatments on the hemagglutinin and its cell receptors. Proc. Soc. exp. Biol. Med. *120:* 127–130 (1965).

261 HEINE, J. W. and SCHNAITMAN, C. A.: Fusion of vesicular stomatitis virus with the cytoplasmic membrane of L cells. J. Virol. *3:* 619–622 (1969).

262 HEINE, J. W. and SCHNAITMAN, C. A.: Entry of vesicular stomatitis virus into L cells. J. Virol *8:* 786–795 (1971).

263 HEINE, J. W.; SPEAR, P. G., and ROIZMAN, B.: Proteins specified by herpes simplex virus. VI. Viral proteins in the plasma membrane. J. Virol. *9:* 431–439 (1972).

264 HENLE, G.; DEINHARDT, F., and GIRARDI, A.: Cytolytic effects of mumps virus in tissue cultures of epithelial cells. Proc. Soc. exp. Biol. Med. *87:* 386–393 (1954).

265 HIRST, G. K.: The agglutination of red cells by allantoic fluid of chick embryos infected with influenza virus. Science *94:* 22–23 (1941).

266 HIRST, G. K.: Adsorption of influenza hemagglutinins and virus by red blood cells. J. exp. Med. *76:* 195–209 (1942).

267 HOCHBERG, E. and BECKER, Y.: Adsorption, penetration and uncoating of herpes simplex virus. J. gen. Virol. *2:* 231–241 (1968).

268 HODGE, L. D. and SCHARFF, M. D.: Effect of adenovirus on host cell DNA synthesis in synchronized cells. Virology *37:* 554–564 (1969).

269 HOFFMANN, C. E.; NEUMAYER, E. M.; HAFF, R. F., and GOLDSBY, R. A.: Mode of action of the antiviral activity of amantadine in tissue culture. J. Bact. *90:* 623–628 (1965).

270 HOLLAND, J. J.: Receptor affinities as major determinants of enterovirus tissue tropisms in humans. Virology *15:* 312–326 (1961).

271 HOLLAND, J. J.: Irreversible eclipse of poliovirus by HeLa cells. Virology *16:* 163–176 (1962).

272 HOLLAND, J. J.: Depression of host-controlled RNA synthesis in human cells during poliovirus infection. Proc. nat. Acad. Sci., Wash. *49:* 23–28 (1963).

273 HOLLAND, J. J.: Enterovirus entrance into specific host cells, and subsequent alterations of cell protein and nucleic acid synthesis. Bact. Rev. *28:* 3–13 (1964).

274 HOLLAND, J. J. and CORDS, C. E.: Maturation of poliovirus RNA with capsid protein coded by heterologous enteroviruses. Proc. nat. Acad. Sci., Wash. *51:* 1082–1085 (1964).

275 HOLLAND, J. J. and HOYER, B. H.: Early stages of enterovirus infection. Cold Spr. Harb. Symp. quant. Biol. *27:* 101–111 (1962).

276 HOLLAND, J. J. and MCLAREN, L. C.: The mammalian cell-virus relationship. II. Adsorption, reception, and eclipse of poliovirus by HeLa cells. J. exp. Med. *109:* 487–504 (1959).

277 HOLLAND, J. J. and MCLAREN, L. C.: The location and nature of enterovirus receptors in susceptible cells. J. exp. Med. *114:* 161–171 (1961).

278 HOLLAND, J. J.; MCLAREN, L. C., and SYVERTON, J. T.: The mammalian cell-virus relationship. IV. Infection of naturally insusceptible cells with enterovirus ribonucleic acid. J. exp. Med. *110:* 65–80 (1959).

279 HOLLAND, J. J. and PETERSON, J. A.: Nucleic acid and protein synthesis during poliovirus infection of human cells. J. molec. Biol. *8:* 556–573 (1964).

280 Holmes, I. H. and Watson, D. H.: An electron microscope study of the attachment and penetration of herpes virus in BHK21 cells. Virology *21:* 112–123 (1961).

281 Holmes, K. V. and Choppin, P. W.: On the role of the response of the cell membrane in determining virus virulence. Contrasting effects of the parainfluenza virus SV5 in two cell types. J. exp. Med. *124:* 501–519 (1966).

282 Holowczak, J. A.: Glycopeptides of vaccinia virus. I. Preliminary characterization and hexosamine content. Virology *42:* 87–99 (1970).

283 Holowczak, J. A.: Uncoating of poxviruses. I. Detection and characterization of subviral particles in the uncoating process. Virology *50:* 216–232 (1972).

284 Holowczak, J. A. and Joklik, W. K.: Studies on the structural proteins of vaccinia virus. I. Structural proteins of virions and cores. Virology *33:* 717–725 (1967).

285 Homma, R.: Reaction of Japanese encephalitis virus with mannan. Acta virol., Praha *12:* 385–396 (1968).

286 Homma, M. and Graham, A. F.: Intracellular fate of Mengo virus ribonucleic acid. J. Bact. *89:* 64–73 (1965).

287 Hosaka, Y.: Biological activities of sonically treated Sendai virus. J. gen. Virol. *8:* 43–54 (1970).

288 Hosaka, Y. and Koshi, Y.: Electron microscopic study of cell fusion by HVJ virions. Virology *34:* 419–434 (1968).

289 Howe, C.; Avrameas, S.; de Vaux St. Cyr, C.; Grabar, P., and Lee, L. T.: Antigenic components of human erythrocytes. J. Immunol. *91:* 683–692 (1963).

290 Howe, C. and Lee, L. T.: Immunochemical study of hemoglobin-free human erythrocyte membranes. J. Immunol. *102:* 573–592 (1969).

291 Howe, C. and Lee, L. T.: Virus-erythrocyte interactions. Adv. Virus Res. *17:* 1–50 (1970).

292 Howe, C. and Morgan, C.: Interactions between Sendai virus and human erythrocytes. J. Virol. *3:* 70–81 (1969).

293 Hoyle, L.: The influenza viruses; in Gard, Hallauer and Meyer Virology Monogr., vol. 4, pp. 1–379 (Springer, Berlin 1968).

294 Hoyle, L.; Horne, R. W., and Waterson, A. P.: The structure and composition of the myxoviruses. III. The interaction of influenza virus particles with cytoplasmic particles derived from normal chorioallantoic membrane cells. Virology *17:* 533–542 (1962).

295 Huang, A. S. and Baltimore, D.: Initiation of polyribosome formation in poliovirus-infected HeLa cells. J. molec. Biol. *47:* 275–291 (1970).

296 Huang, A. S.; Baltimore, D., and Bratt, M. A.: Ribonucleic acid polymerase in virions of Newcastle disease virus. Comparison with the vesicular stomatitis virus polymerase. J. Virol. *7:* 389–394 (1971).

297 Huang, A. S. and Wagner, R. R.: Penetration of herpes simplex virus into human epidermoid cells. Proc. Soc. exp. Biol. Med. *116:* 863–869 (1964).

298 Huang, A. S. and Wagner, R. R.: Inhibition of cellular RNA synthesis by nonreplicating vesicular stomatitis virus. Proc. nat. Acad. Sci., Wash. *54:* 1579–1584 (1965).

299 Huang, A. S. and Wagner, R. R.: Defective T particles of vesicular stomatitis virus. II. Biologic role in homologous interference. Virology *30:* 173–181 (1966).

299a Huang, E. S.; Estes, M. K., and Pagano, J. S.: Structure and function of the polypeptides in simian virus 40. I. Existence of subviral deoxynucleoprotein complexes. J. Virol. *9:* 923–929 (1972).

300 Huang, E. S.; Nonoyama, M., and Pagano, J. S.: Structure and function of the polypeptides in simian virus 40. II. Transcription of subviral deoxynucleoprotein complexes in vitro. J. Virol. *9:* 930–937 (1972).

301 Hughes, R. C.: Glycoproteins as components of cellular membranes; in Butler and Noble Progress in biophysics and molecular biology, vol. 26, pp. 189–268 (Pergamon Press, New York 1973).

302 Hughes, R. C.; Sanford, B., and Jeanloz, R. W.: Regeneration of the surface glycoproteins of a transplantable mouse tumor cell after treatment with neuraminidase. Proc. nat. Acad. Sci., Wash. *69:* 942–945 (1972).

303 Hultin, J. V. and McKee, A. P.: Fixation of 'neutralized' influenza virus by susceptible cells. J. Bact. *63:* 437–447 (1952).

304 Hummeler, K.; Koprowski, H., and Wiktor, T. J.: Structure and development of rabies virus in tissue culture. J. Virol. *1:* 152–170 (1967).

305 Hummeler, K.; Tomassini, N., and Sokol, F.: Morphological aspects of the uptake of SV40 by permissive cells. J. Virol. *6:* 87–93 (1970).

306 Hummeler, K.; Tomassini, N., and Zajac, B.: Early events in herpes simplex virus infection. A radioautographic study. J. Virol. *4:* 67–74 (1969).

307 Inbar, M.; Ben-Bassat, H., and Sachs, L.: A specific metabolic activity on the surface membrane in malignant cell-transformation. Proc. nat. Acad. Sci., Wash. *68:* 2748–2751 (1971).

308 Incardona, N. L.; Blonski, R., and Feeney, W.: Mechanism of adsorption and eclipse of bacteriophage ØX174. I. In vitro conformational change under conditions of eclipse. J. Virol. *9:* 96–101 (1972).

309 Inglot, A. D.; Kochman, M., and Mastalerz, P.: Activity of salicylates as inhibitors of virus replication in vitro and their mode of action. Acta virol., Praha *10:* 185–194 (1966).

310 Ishibashi, M. and Maizel, J. V.: Hexosamine in fiber protein of type 2 adenovirus. Abstracts 72nd Ann. Meet. Amer. Soc. Microbiol., 1972, p. 199 (v. 86).

311 Ishida, N. and Ackermann, W. W.: Growth characteristics of influenza virus. Properties of the initial cell-virus complex. J. exp. Med. *104:* 505–515 (1956).

312 Jacobson, M. F.; Asso, J., and Baltimore, D.: Further evidence of the formation of poliovirus proteins. J. molec. Biol. *49:* 657–669 (1970).

313 Jacobson, M. F. and Baltimore, D.: Morphogenesis of poliovirus. I. Association of the viral RNA with coat protein. J. molec. Biol. *33:* 369–378 (1968).

314 Jirgensons, B. and Springer, G. F.: Conformation of blood group and virus receptor glycoproteins from red cells and secretions. Science *162:* 365–367 (1968).

315 Johnson, T. C. and Holland, J. J.: Ribonucleic acid and protein synthesis in mitotic HeLa cells. J. Cell Biol. *27:* 565–574 (1965).

316 Joklik, W. K.: The intracellular uncoating of poxvirus DNA. I. The fate of radioactively-labeled rabbitpox virus. J. molec. Biol. *8:* 263–276 (1964).

317 JOKLIK, W. K.: The intracellular uncoating of poxvirus DNA. II. The molecular basis of the uncoating process. J. molec. Biol. *8:* 277–288 (1964).

318 JOKLIK, W. K.: The molecular basis of the viral eclipse phase; in MELNICK Progr. med. Virol., vol. 7, pp. 44–96 (Karger, Basel 1965).

319 JOKLIK, W. K.: The poxviruses. Bact. Rev. *30:* 33–66 (1966).

320 JOKLIK, W. K.: Studies on the effect of chymotrypsin on reovirions. Virology *49:* 700–715 (1972).

321 JOKLIK, W. K. and DARNELL, J. E.: The adsorption and early fate of purified poliovirus in HeLa cells. Virology *13:* 439–447 (1961).

322 JOKLIK, W. K. and MERIGAN, T. C.: Concerning the mechanism of action of interferon. Proc. nat. Acad. Sci., Wash. *56:* 558–565 (1966).

323 JOKLIK, W. K. and ZWEERINK, H. J.: The morphogenesis of animal viruses. Ann. Rev. Genet. *5:* 297–360 (1971).

324 JUNGEBLUT, C. W. and KODZA, H.: Studies on propagation of the Col SK group of viruses in various tissue culture media. Proc. Soc. exp. Biol. Med. *96:* 133–139 (1957).

325 JUNGWIRTH, C. and LAUNER, J.: Effect of poxvirus infection on host cell deoxyribonucleic acid synthesis. J. Virol. *2:* 401–408 (1968).

326 KABAT, E. A.: Size and heterogeneity of the combining sites on an antibody molecule. J. Cell Physiol. *50:* suppl. 1, pp. 79–102 (1957).

327 KANG, C. Y. and PREVEC, L.: Proteins of vesicular stomatitis virus. II. Immunological comparisons of viral antigens. J. Virol. *6:* 20–27 (1970).

328 KAPLAN, A. S. and BEN-PORAT, T.: Synthesis of proteins in cells infected with herpesvirus. VI. Characterization of the proteins of the viral membrane. Proc. nat. Acad. Sci., Wash. *66:* 799–806 (1970).

329 KAPLAN, M. M.; WIKTOR, T. J.; MAES, R. F.; CAMPBELL, J. B., and KOPROWSKI, H.: Effect of polyions on the infectivity of rabies virus in tissue culture. Construction of a single-cycle growth curve. J. Virol. *1:* 145–151 (1967).

330 KASAMAKI, A.; BEN-PORAT, T., and KAPLAN, A. S.: Polyoma virus-induced release of inhibition of cellular DNA synthesis caused by iododeoxyuridine. Nature, Lond. *217:* 756–758 (1968).

331 KASEL, J. A. and HUBER, M.: Relationship of the adenovirus erythrocyte-receptor-modifying factor to the type-specific complement-fixing antigen. Proc. Soc. exp. Biol. Med. *116:* 16–18 (1964).

332 KASEL, J. A.; ROWE, W. P., and NEMES, J. L.: Further characterization of the adenovirus erythrocyte receptor-modifying factor. J. exp. Med. *114:* 717–728 (1961).

333 KATAGIRI, S.; AIKAWA, S., and HINUMA, Y.: Stepwise degradation of poliovirus capsid by alkaline treatment. J. gen. Virol. *13:* 101–109 (1971).

334 KATAGIRI, S.; HINUMA, Y., and ISHIDA, N.: Relation between the adsorption to cells and antigenic properties in poliovirus particles. Virology *34:* 797–799 (1968).

335 KATHAN, R. H.; RIFF, L. J. M., and REAL, M.: Association between the erythrocyte hemagglutination inhibitor and the M-N blood group substances. Proc. Soc. exp. Biol. Med. *114:* 90–92 (1963).

336 KATHAN, R. H.; WINZLER, R. J., and JOHNSON, C. A.: Preparation of an inhibitor of viral hemagglutination from human erythrocytes. J. exp. Med. *113:* 37–45 (1961).

337 KATO, N. and EGGERS, H. J.: Inhibition of uncoating of fowl plague virus by l-adamantanamine hydrochloride. Virology *37:* 632–641 (1969).

338 KELLER, R.: Studies on the mechanism of the enzymatic reactivation of antibody-neutralized poliovirus. J. Immunol. *100:* 1071–1079 (1968).

339 KERN, J. and ROSEN, L.: Factors affecting hemagglutination by enteroviruses. Proc. Soc. exp. Biol. Med. *115:* 536–541 (1964).

340 KERN, J. and ROSEN, L.: Identification of enteroviruses by hemagglutination-inhibition. J. Bact. *91:* 1936–1942 (1966).

341 KERR, I. M.; MARTIN, E. M.; HAMILTON, M. G., and WORK, T. S.: The initiation of virus protein synthesis in Krebs ascites tumor cells infected with EMC virus. Cold Spr. Harb. Symp. quant. Biol. *27:* 259–269 (1962).

342 KHARE, G. P. and CONSIGLI, R. A.: Multiplication of polyoma virus. I. Use of selectively labeled ($H^3$) virus to follow the course of infection. J. Bact. *90:* 819–821 (1965).

343 KHESIN, Y. E. and AMCHENKOVA, A. M.: On the relationship between the enzymatic composition and the susceptibility of the cell to viruses. J. gen. Virol. *15:* 105–111 (1972).

344 KILBOURNE, E. D.; CHOPPIN, P. W.; SCHULZE, I. T.; SCHOLTISSEK, C., and BUCHER, D. L.: Influenza virus polypeptides and antigens. Summary of influenza workshop 1. J. infect. Dis. *125:* 447–455 (1972).

345 KIT, S. and DUBBS, D. R.: Biochemistry of vaccinia-infected mouse fibroblasts (strain L-M). II. Properties of the chromosomal DNA of infected cells. Virology *18:* 286–293 (1962).

346 KLENK, H. D. and CHOPPIN, P. W.: Glycosphingolipids of plasma membranes of cultured cells and an enveloped virus (SV5) grown in these cells. Proc. nat. Acad. Sci., Wash. *66:* 57–64 (1970).

347 KLENK, H. D. and CHOPPIN, P. W.: Plasma membrane lipids and parainfluenza virus assembly. Virology *40:* 939–947 (1970).

348 KLENK, H. D. and CHOPPIN, P. W.: Glycolipid content of vesicular stomatitis virus grown in baby hamster kidney cells. J. Virol. *7:* 416–417 (1971).

349 KLENK, H. D.; COMPANS, R. W., and CHOPPIN, P. W.: An electron microscopic study of the presence or absence of neuraminic acid in enveloped viruses. Virology *42:* 1158–1162 (1970).

350 KLENK, H. D.; ROTT, R., and BECHT, H.: On the structure of the influenza virus envelope. Virology *47:* 579–591 (1972).

351 KLUG, A.: Design of self-assembling systems of equal units; in WARREN Formation and fate of cell organelles, pp. 1–18 (Academic Press, New York 1967).

352 KLUG, A.; FINCH, J. T.; LEBERMAN, R., and LONGLEY, W.: Design and structure of regular virus particles. Ciba Found. Symp. 1965; in WOLSTENHOLME and O'CONNOR Principles of biomolecular organization, pp. 158–191 (Churchill, London 1965).

353 KNIPPERS, R.; SALIVAR, W. O.; NEWBOLD, J. E., and SINSHEIMER, R. L.: The process of infection with bacteriophage ØX174. XXVI. Transfer of the parental DNA of bacteriophage ØX174 into progeny bacteriophage particles. J. molec. Biol. *39:* 641–654 (1969).

354 KOCH, A. and GYÖRGY, E.: Studies on the initial phases of poliovirus reproduction cycle. Acta microbiol. hung. *13:* 85–94 (1966).

355 KOCH, A. and GYÖRGY, E.: Heart glycosides in poliovirus host cell interaction. I. Effect of digoxin and digitoxin and their aglucons on one step growth curves. Acta microbiol. hung. *16:* 189–196 (1969).

356 KOCH, A. and GYÖRGY, E.: Heart glycosides in poliovirus host cell interaction. III. Chemical structure and activity. Acta microbiol. hung. *16:* 245–251 (1969).

357 KOCH, A. and GYÖRGY, E.: Action of cation transfer ATPase inhibitors on efficiency of infection with polioviruses. Acta microbiol. hung. *17:* 127–131 (1970).

358 KOCH, A.; GYÖRGY, E.; NAFICY, A. R., and CSEH, J.: Studies on the possibly digitoxin-sensitive phase of poliovirus-host cell interaction. Acta virol., Praha *14:* 343–352 (1970).

359 KOCH, A. L.: Encounter efficiency of coliphage-bacterium interaction. Biochim. biophys. Acta *39:* 311–318 (1960).

360 KOCH, A. S. and FEHÉR, G.: The possible nature of the chance-event in initiation of virus infection at the cellular level. J. gen. Virol. *18:* 319–327 (1973).

361 KODZA, H. and JUNGEBLUT, C. W.: Effect of receptor-destroying enzyme on the growth of EMC virus in tissue culture. J. Immunol. *81:* 76–81 (1958).

362 KORANT, B. D.; LONBERG-HOLM, K.; NOBLE, J., and STASNY, J. T.: Naturally occurring and artificially produced components of three rhinoviruses. Virology *48:* 71–86 (1972).

363 KOSYAKOV, P. N.; ROVNOVA, Z. I., and SCHACHAVELOYOVA, A. P.: Neutralization of myxoviruses by anticellular sera. Acta virol., Praha *10:* 218–225 (1966).

364 KRIŽANOVÁ, O. and ČIAMPOR, F.: Interaction of plasma membranes with influenza virus. I. Isolation of plasma membranes from chick embryo cells. Acta virol., Praha *15:* 19–27 (1971).

365 KRIŽANOVÁ, O.; KOČISKOVÁ, D.; RATHOVÁ, V., and STYK, B.: Interaction of plasma membranes with influenza virus. II. Release of viral nucleoprotein. Acta virol., Praha *15:* 352–360 (1971).

366 KUNIN, C. M.: Virus-tissue union and the pathogenesis of enterovirus infections. J. Immunol. *88:* 556–569 (1962).

367 KUNIN, C. M.: Cellular susceptibility to enteroviruses. Bact. Rev. *28:* 382–390 (1964).

368 KUNIN, C. M.: Distribution of cell receptors and simple sugar inhibitors during encephalomyocarditis virus-cell union. J. Virol. *1:* 274–282 (1967).

369 KUNIN, C. M. and HALMAGYI, N. E.: The relative abundance of viral receptors. An explanation of the differential susceptibility of suckling and adult mice to coxsackie $B_1$ infection. J. clin. Invest. *40:* 1055–1056 (1961).

370 KUSANO, T.; WANG, R.; POLLACK, R., and GREEN, H.: Human-mouse hybrid cell lines and susceptibility to poliovirus. II. Polio sensitivity and the chromosome constitution of the hybrids. J. Virol. *5:* 682–685 (1970).

371 Kuwert, E.; Wiktor, T. J.; Sokol, F., and Koprowski, H.: Hemagglutination by rabies virus. J. Virol. *2:* 1381–1392 (1968).

372 La Colla, P.; Marcialis, M. A.; Mereu, G. P., and Loddo, B.: Specific inhibition of poliovirus induced blockade of cell protein synthesis by a thiopyrimidine derivative. J. gen. Virol. *17:* 13–18 (1972).

373 Lacour, F.; Fourcade, A.; Verger, C., and Delaine, E.: Coiled structure of the nucleocapsid of avian myeloblastosis virus. J. gen. Virol. *9:* 89–92 (1970).

374 Lafferty, K. J.: The interaction between virus and antibody. I. Kinetic studies. Virology *21:* 61–75 (1963).

375 Lahelle, O. and Horsfall, F. C.: Hemagglutination with the GDVII strain of mouse encephalomyelitis virus. Proc. Soc. exp. Biol. Med. *71:* 713–718 (1949).

376 Lai, M. M. C. and Duesberg, P. H.: Differences between the envelope glycoproteins and glycopeptides of avian tumor viruses released from transformed and nontransformed cells. Virology *50:* 359–372 (1972).

377 Landau, B. J.; Crowell, R. L.; Boclair, C. W., and Zajac, B. A.: The permissiveness of differentiating mouse muscle-cell cultures to infection by group A coxsackieviruses types 1 and 5. Proc. Soc. exp. Biol. Med. *141:* 753–758 (1972).

378 Lauffer, M. A. and Stevens, C. L.: Structure of the tobacco mosaic virus particle: polymerization of tobacco mosaic virus protein. Adv. Virus Res. *13:* 1–63 (1968).

379 Laver, W. G.; Pereira, W. C., and Valentine, R. C.: Isolation of an internal component from adenovirus type 5. J. molec. Biol. *37:* 379–386 (1968).

380 Laver, W. G. and Valentine, R. C.: Morphology of the isolated hemagglutinin and neuraminidase subunits of influenza virus. Virology *38:* 105–119 (1969).

381 Laver, W. G.; Wrigley, N. G., and Pereira, H. G.: Removal of pentons from particles of adenovirus type 2. Virology *39:* 599–605 (1969).

382 Law, J. H. and Snyder, W. R.: Membrane lipids; in Fox and Keith Membrane molecular biology, pp. 3–26 (Sinauer Associates, Stamford 1972).

383 Lawrence, W. C. and Ginsberg, H. S.: Intracellular uncoating of type 5 adenovirus deoxyribonucleic acid. J. Virol. *1:* 851–867 (1967).

384 Ledinko, N. and Fong, C. K. Y.: Kinetics of nucleic acid synthesis in human embryonic kidney cultures infected with adenovirus 2 or 12: inhibition of cellular deoxyribonucleic acid synthesis. J. Virol. *4:* 123–132 (1969).

385 Leibowitz, R. and Penman, S.: Regulation of protein synthesis in HeLa cells. III. Inhibition during poliovirus infection. J. Virol. *8:* 661–668 (1971).

386 Lerner, A. M.; Bailey, E. J., and Kofender, M.: Preparations of saliva inhibiting reovirus hemagglutination. J. Immunol. *96:* 59–63 (1966).

387 Lerner, A. M.; Bailey, E. J., and Tillotson, J. R.: Enterovirus hemagglutination. Inhibition by several enzymes and sugars. J. Immunol. *95:* 1111–1115 (1965).

388 Lerner, A. M.; Cherry, J. D., and Finland, M.: Hemagglutination with reoviruses. Virology *19:* 58–65 (1963).

389 Lerner, A. M.; Gelb, L. D.; Tillotson, J. R.; Carruthers, M. M., and Bailey, E. J.: Enterovirus hemagglutination. Inhibition by aldoses and a possible mechanism. J. Immunol. *96:* 629–636 (1966).

390 Lerner, A. M. and Miranda, Q. R.: Cellular interactions of several enteroviruses and a reovirus after treatment with sodium borohydride or carbohydrases. Virology *36:* 277–285 (1968).

391 Levine, A. J. and Ginsberg, H. S.: Biochemical studies on the mechanism by which the fiber antigen inhibits multiplication of type 5 adenovirus. J. Virol. *1:* 447–457 (1967).

392 Levine, A. J. and Ginsberg, H. S.: Role of adenovirus structural proteins in the cessation of host-cell biosynthetic functions. J. Virol. *2:* 430–439 (1968).

393 Levinthal, J.; Dunnebacke, T. H., and Williams, R. C.: Indirect immunoferritin study of poliovirus-, Sendai- and Sindbis-infected cells. J. Ultrastruct. Res. *30:* 244 (1970).

394 Levitt, N. H. and Crowell, R. L.: Comparative studies of the regeneration of HeLa cell receptors for poliovirus T1 and coxsackievirus B3. J. Virol. 1: 693–700 (1967).

395 Lewandowski, L. J. and Traynor, B. L.: Comparison of the structure and polypeptide composition of three double-stranded ribonucleic acid-containing viruses (diplornaviruses). Cytoplasmic polyhedrosis virus, wound tumor virus, and reovirus. J. Virol. *10:* 1053–1070 (1972).

395a Lewenton-Kriss, S. and Mandel, B.: Studies on the nonneutralizable fraction of poliovirus. Virology *48:* 666–678 (1972).

396 Liebhaber, H. and Takemoto, K. K.: The basis for the size differences in plaques produced by variants of encephalomyocarditis (EMC) virus. Virology *20:* 559–566 (1963).

397 Lilly, F.: The inheritance of susceptibility to the Gross leukemia virus in mice. Genetics *53:* 529–539 (1966).

398 Lin, H.: Inhibition of thymidine kinase activity and deoxyribonucleic acid synthesis in L cells infected with meningopneumonitis agent. J. Bact. *96:* 2054–2063 (1968).

399 Lonberg-Holm, K. and Korant, B. D.: Early interaction of rhinoviruses with host cells. J. Virol. *9:* 29–40 (1972).

400 Lonberg-Holm, K. and Noble-Harvey, J.: Comparison of *in vitro* and cell-mediated alteration of a human rhinovirus and its inhibition by SDS. J. Virol. *12:* 819–826 (1973).

401 Lonberg-Holm, K. and Philipson, L.: Early events of virus-cell interaction in an adenovirus system. J. Virol. *4:* 323–338 (1969).

402 Lonberg-Holm, K. and Yin, F. H.: Antigenic determinants of infective and inactivated human rhinovirus type 2. J. Virol. *12:* 114–123 (1973).

402a Lucy, J. A.: The fusion of biological membranes. Nature, Lond. *227:* 815–817 (1970).

403 Luftig, R. B. and Kilham, S. S.: An electron microscope study of Rauscher leukemia virus. Virology 46: 277–297 (1971).

404 Luftig, R. B.; Kilham, S. S.; Hay, A. J.; Zweerink, H. J., and Joklik, W. K.: An ultrastructural study of virions and cores of reovirus type 3. Virology *48:* 170–181 (1972).

405 Lukert, P. D.: Chemical characterization of avian infectious bronchitis virus receptor sites. Amer. J. vet. Res. *33:* 987–994 (1972).

406 Luria, S. E.: Méthodes statistiques appliquées à l'étude du mode d'action des ultravirus. Ann. Inst. Pasteur *64:* 415–438 (1940).

407 LURIA, S. E.: Bacteriophage. An essay on virus reproduction. Science *111:* 507–511 (1950).

408 LWOFF, A.; HORNE, R., and TOURNIER, P.: A system of viruses. Cold Spr. Harb. Symp. quant. Biol. *27:* 51–55 (1962).

409 MADDY, A. H.: The solubilization of the protein of the Ox-erythrocyte ghost. Biochim. biophys. Acta *88:* 448–449 (1964).

410 MADEN, B. E. H.; VAUGHAN, M. H.; WARNER, J. R., and DARNELL, J. E.: Effects of valine deprivation on ribosome formation in HeLa cells. J. molec. Biol. *45:* 265–275 (1969).

411 MAIZEL, J. V.; PHILLIPS, B. A., and SUMMERS, D. F.: Composition of artificially produced and naturally occurring empty capsids of poliovirus type 1. Virology *32:* 692–699 (1967).

412 MAIZEL, J. V.; WHITE, D. O., and SCHARFF, M. D.: The polypeptides of adenovirus. II. Soluble proteins, cores, top components and the structure of the virion. Virology *36:* 126–136 (1968).

413 MAK, T. W.; O'CALLAGHAN, D. J., and COLTER, J. S.: Studies of the pH inactivation of three variants of Mengo encephalomyelitis virus. Virology *40:* 565–571 (1970).

414 MAK, T. W.; O'CALLAGHAN, D. J., and COLTER, J. S.: Studies of the early events of the replicative cycle of three variants of Mengo encephalomyelitis virus in mouse fibroblast cells. Virology *42:* 1087–1096 (1970).

415 MAK, T. W.; O'CALLAGHAN, D. J.; KAY, C. M., and COLTER, J. S.: Studies of the protein subunit of pH-inactivated Mengo virus variants. II. Physicochemical properties. Virology *43:* 579–587 (1971).

416 MÄKELÄ, O.; MIETTINEN, T., and PESOLA, R.: Release of sialic acid and carbohydrates from human red cells by trypsin treatment. Vox Sang. *5:* 492–496 (1960).

417 MANDEL, B.: Studies on the interactions of poliomyelitis virus, antibody, and host cells in a tissue culture system. Virology *6:* 424–447 (1958).

418 MANDEL, B.: Reversibility of the reaction between poliovirus and neutralizing antibody of rabbit origin. Virology *14:* 316–328 (1961).

419 MANDEL, B.: Early stages of virus-cell interaction as studied by using antibody. Cold Spr. Harb. Symp. quant. Biol. *27:* 123–131 (1962).

420 MANDEL, B.: The use of sodium dodecyl sulfate in studies on the interaction of poliovirus and HeLa cells. Virology *17:* 288–294 (1962).

420a MANDEL, B.: The fate of the inoculum in HeLa cells infected with poliovirus. Virology *25:* 152–154 (1965).

421 MANDEL, B.: The interaction of neutralized poliovirus with HeLa cells. I. Adsorption. Virology *31:* 238–247 (1967).

422 MANDEL, B.: The interaction of neutralized poliovirus with HeLa cells. II. Elution, penetration, uncoating. Virology *31:* 248–259 (1967).

423 MANDEL, B.: The relationship between penetration and uncoating of poliovirus in HeLa cells. Virology *31:* 702–712 (1967).

424 MANDEL, B.: Characterization of type 1 poliovirus by electrophoretic analysis. Virology *44:* 554–568 (1971).

425 MANDEL, B.: An analysis of the physical and chemical factors involved in the reactivation of neutralized poliovirus by the method of freezing and thawing. Virology *51:* 358–369 (1973).

426 MANDEL, B. and RACKER, E.: Inhibition of Theiler's encephalomyelitis virus (GDVII strain) of mice by an intestinal mucopolysaccharide. J. exp. Med. *98:* 399–416 (1953).

427 MANDERS, E. K.; TILLES, J. G., and HUANG, A. S.: Interferon-mediated inhibition of virion-directed transcription. Virology *49:* 573–581 (1972).

428 MANN, J. J.; ROSSEN, R. D.; LEHRICH, J. R., and KASEL, J. A.: The effect of kaolin on immunoglobulins. An improved technique to remove the nonspecific serum inhibitor of reovirus hemagglutination. J. Immunol. *98:* 1136–1142 (1967).

429 MARCHESI, V. T. and ANDREWS, E. P.: Glycoproteins. Isolation from cell membranes with lithium diiodosalicylate. Science *174:* 1247–1248 (1971).

430 MARCHESI, V. T.; TILLACK, T. W.; JACKSON, R. L.; SEGREST, J. P., and SCOTT, R. E.: Chemical characterization and surface orientation of the major glycoprotein of the human erythrocyte membrane. Proc. nat. Acad. Sci., Wash. *69:* 1445–1449 (1972).

431 MARCHESI, V. T.; TILLACK, T. W., and SCOTT, R. E.: The role of glycoproteins in red cell membrane structure; in JAMIESON and GREENWALT Glycoproteins of blood cells and plasma, pp. 94–105 (Lippincott, Philadelphia 1971).

432 MARCUS, P. I.; ENGELHARDT, D. L.; HUNT, J. M., and SEKELLICK, M. J.: Interferon action. Inhibition of vesicular stomatitis virus RNA synthesis induced by virion-bound polymerase. Science *174:* 593–598 (1971).

433 MARCUS, P. I. and HIRSH, D.: Renewal of a normal surface component. Cellular receptors for virus attachment. J. Cell Biol. *19:* 47A (1963).

434 MARTIN, S. A. and ZWEERINK, H. J.: Isolation and characterization of two types of bluetongue virus particles. Virology *50:* 495–506 (1972).

435 MARTIN, E. M.; MALEC, J.; SVED, S., and WORK, T. S.: Studies on protein and nucleic acid metabolism in virus-infected mammalian cells. I. Encephalomyocarditis virus in Krebs II mouse-ascites-tumor cells. Biochem. J. *80:* 585–597 (1961).

436 MARTIN, E. M. and WORK, T. S.: Studies on protein and nucleic acid metabolism in virus-infected mammalian cells. 5. The kinetics of synthesis of virus protein and of virus ribonucleic acid in Krebs II mouse-ascites-tumour cells infected with encephalomyocarditis virus. Biochem. J. *83:* 574–582 (1962).

437 MATTERN, C. F. T.: Virus architecture as determined by X-ray diffraction and electron microscopy; in LEVY The biochemistry of viruses, pp. 55–100 (Marcel Dekker, New York 1969).

438 MATTERN, C. F. T.; TAKEMOTO, K. K., and DANIEL, W. A.: Replication of polyoma virus in mouse embryo cells. Electron microscopic observations. Virology *30:* 242–256 (1966).

439 MAYER, M. M.; RAPP, H. J.; ROIZMAN, B.; KLEIN, S. W.; COWAN, K. M.; LUKENS, D.; SCHWERDT, C. E.; SCHAFFER, F. L., and CHARNEY, J.: The purification of poliomyelitis virus as studied by complement fixation. J. Immunol. *78:* 435–455 (1957).

440 McClelland, L. and Hare, R.: The adsorption of influenza virus by red cells and a new in vitro method of measuring antibodies for influenza virus. Canad. Publ. Hlth J. *32:* 530–538 (1941).

441 McCormick, W. and Penman, S.: Inhibition of RNA synthesis in HeLa and L cells by Mengovirus. Virology *31:* 135–141 (1967).

442 McLaren, L. C.; Holland, J. J., and Syverton, J. T.: The mammalian cell-virus relationship. I. Attachment of poliovirus to cultivated cells of primate and non-primate origin. J. exp. Med. *109:* 475–485 (1959).

443 McLaren, L. C.; Holland, J. J., and Syverton, J. T.: The mammalian cell-virus relationship. V. Susceptibility and resistance of cells *in vitro* to infection by Coxsackie A9 virus. J. exp. Med. *112:* 581–594 (1960).

444 McLaren, L. C.; Scaletti, J. V., and James, C. G.: Isolation and properties of enterovirus receptors; in Manson Biological properties of the mammalian surface membrane, pp. 123–135 (Wistar Institute Press, Philadelphia 1968).

445 McSharry, J. J. and Wagner, R. R.: Lipid composition of purified vesicular stomatitis viruses. J. Virol. *7:* 59–70 (1971).

446 Medrano, L. and Green, H.: Picornavirus receptors and picornavirus multiplication in human mouse hybrid cell lines. Virology *54:* 515–525 (1973).

447 Medzon, E. L. and Bauer, H.: Structural features of vaccinia virus revealed by negative staining, sectioning, and freeze-etching. Virology *40:* 860–867 (1970).

448 Meezan, E.; Wu, H. C.; Black, P. H., and Robbins, P. W.: Comparative studies on the carbohydrate-containing membrane components of normal and virus-transformed mouse fibroblasts. II. Separations of glycoproteins and glycopeptides by Sephadex chromatography. Biochemistry *8:* 2518–2524 (1969).

449 Meiselman, N.; Kohn, A., and Danon, D.: Electron microscopic study of penetration of Newcastle disease virus into cells leading to formation of polykaryocytes. J. Cell Sci. *2:* 71–76 (1967).

450 Miller, G. and Enders, J. F.: Vaccinia virus replication and cytopathic effect in cultures of phytohemagglutinin-treated human peripheral blood leukocytes. J. Virol. *2:* 787–792 (1968).

451 Miyamoto, K. and Gilden, R. V.: Electron microscopic studies of tumor viruses. I. Entry of murine leukemia virus into mouse embryo fibroblasts. J. Virol. *7:* 395–406 (1971).

452 Miyamoto, K. and Morgan, C.: Structure and development of viruses as observed in the electron microscope. XI. Entry and uncoating of herpes simplex virus. J. Virol. *8:* 910–918 (1971).

453 Moolten, F. L. and Kibrick, S.: Enhancement of Sindbis virus infectivity by reduced salt concentration. Proc. Soc. exp. Biol. Med. *136:* 335–340 (1971).

454 Morawiecki, A. and Lisowska, E.: Polymerized orosomucoid – an inhibitor of influenza virus hemagglutination. Biochem. biophys. Res. Commun. *18:* 606–610 (1965).

455 Morgan, C.: Entry of reovirus. Bact. Proc. *1970:* 203 (V322).

456 Morgan, C. and Howe, C.: Structure and development of viruses as observed in the electron microscope. IX. Entry of parainfluenza I (Sendai) virus. J. Virol. *2:* 1122–1132 (1968).

457 Morgan, C. and Rose, H. M.: Structure and development of viruses as observed in the electron microscope. VIII. Entry of influenza virus. J. Virol. *2:* 925–936 (1968).

458 Morgan, C.; Rose, H. M., and Mednis, B.: Electron microscopy of herpes simplex virus. I. Entry. J. Virol. *2:* 507–516 (1968).

459 Morgan, C.; Rosenkranz, H. S., and Mednis, B.: Structure and development of viruses as observed in the electron microscope. X. Entry and uncoating of adenovirus. J. Virol. *4:* 777–796 (1969).

460 Morgan, H. R.; Enders, J. F., and Wagley, P. F.: A hemolysin associated with the mumps virus. J. exp. Med. *88:* 503–514 (1948).

461 Mori, H.; Nakai, H., and Nozima, T.: The interaction of Japanese encephalitis virus with its receptor prepared from pigeon red blood cells and susceptible cells. Acta virol., Praha *9:* 97–108 (1965).

462 Mori, R.; Schieble, J. H., and Ackermann, W. W.: Reaction of polyoma and influenza viruses with receptors of erythrocytes and host cells. Proc. Soc. exp. Biol. Med. *109:* 685–690 (1962).

463 Moss, B.: Inhibition of HeLa cell protein synthesis by the vaccinia virion. J. Virol. *2:* 1028–1037 (1968).

464 Mountcastle, W. E.; Compans, R. W.; Caliguiri, L. A., and Choppin, P. W.: Nucleocapsid protein subunits of simian virus 5, Newcastle disease virus, and Sendai virus. J. Virol. *6:* 677–684 (1970).

465 Mountcastle, W. E.; Compans, R. W., and Choppin, P. W.: Proteins and glycoproteins of paramyxoviruses. A comparison of simian virus 5, Newcastle disease virus, and Sendai virus. J. Virol. *7:* 47–52 (1971).

466 Much, D. H. and Zajac, I.: Homology of surface receptors for poliovirus on mammalian cell lines. J. gen. Virol. *21:* 385–390 (1973).

467 Mudd, J. A. and Summers, D. F.: Protein synthesis in vesicular stomatitis virus-infected HeLa cells. Virology *42:* 328–340 (1970).

468 Nahmias, A. J. and Kibrick, S.: Inhibitory effect of heparin on herpes simplex virus. J. Bact. *87:* 1060–1066 (1964).

469 Nahmias, A. J.; Kibrick, S., and Rosan, R. C.: Viral leukocyte interrelationships. I. Multiplication of a DNA virus – herpes simplex – in human leukocyte cultures. J. Immunol. *93:* 69–74 (1964).

470 Nakai, H.; Homma, R., and Nozima, T.: Interaction of Japanese encephalitis virus with mouse liver cells and with the receptor extracted therefrom. Acta virol., Praha *10:* 400–405 (1966).

471 Nardone, R. M.; Todd, J.; Gonzalez, P., and Gaffney, E. V.: Nucleoside incorporation into strain L cells: inhibition by pleuropneumonia-like organisms. Science *149:* 1100–1101 (1965).

472 Neff, J. M. and Enders, J. F.: Poliovirus replication and cytopathogenicity in monolayer hamster cell cultures fused with beta propiolactone-inactivated Sendai virus. Proc. Soc. exp. Biol. Med. *127:* 260–267 (1968).

473 Nermut, M. V.; Frank, H., and Schäfer, W.: Properties of mouse leukemia viruses. III. Electron microscopic appearance as revealed after conventional preparation techniques as well as freeze-drying and freeze-etching. Virology *49:* 345–358 (1972).

474 NEUMAYER, E. M.; HAFF, R. F., and HOFFMANN, C. E.: Antiviral activity of amantadine hydrochloride in tissue culture and *in vivo*. Proc. Soc. exp. Biol. Med. *119:* 393–396 (1965).

475 NEURATH, A. R.: Changes in the haemolytic activity of Sendai virus after different chemical and physical treatments. Acta virol., Praha *8:* 143–153 (1964).

476 NEURATH, A. R.; HARTZELL, R. W., and RUBIN, B. A.: Solubilization and some properties of erythrocyte receptors for adenovirus type 7 hemagglutinin. Nature, Lond. *221:* 1069–1071 (1969).

477 NEURATH, A. R.; HARTZELL, R. W., and RUBIN, B. A.: Partial characterization of the complementary sites involved in the reaction between adenovirus type 7 and erythrocyte receptors. Virology *42:* 789–793 (1970).

478 NEURATH, A. R. and NORRBY, E. C. J.: Further elucidation of the nature of the measles virus hemolysin based on inactivation studies. Arch. ges. Virusforsch. *15:* 651–658 (1965).

479 NEWBOLD, J. E. and SINSHEIMER, R. L.: The process of infection with bacteriophage ØX174. XXXII. Early steps in the infection process. Attachment, eclipse and DNA penetration. J. molec. Biol. *49:* 49–66 (1970).

480 NEWBOLD, J. E. and SINSHEIMER, R. L.: Process of infection with bacteriophage ØX174. XXXIV. Kinetics of the attachment and eclipse steps of the infection. J. Virol. *5:* 427–431 (1970).

481 NEWTON, A.: In MELNICK Int. Virol., vol. 1, p. 253 (Karger, Basel 1968).

482 NICOLSON, G. L.: Difference in topology of normal and tumour cell membranes shown by different surface distributions of ferritin-conjugated concanavalin A. Nature new Biol. *233:* 244–246 (1971).

483 NICOLSON, G. L.: Topography of membrane concanavalin A sites modified by proteolysis. Nature new Biol. *239:* 193–197 (1972).

484 NICOLSON, G. L.: Temperature-dependent mobility of concanavalin A sites on tumor cell surface. Nature new Biol. *243:* 218–220 (1973).

485 NICOLSON, G. L. and SINGER, S. J.: Ferritin-conjugated plant agglutinins as specific saccharide stains for electron microscopy. Application to saccharides bound to cell membranes. Proc. nat. Acad. Sci., Wash. *68:* 942–945 (1971).

486 NII, S.; MORGAN, C.; ROSE, H. M., and HSU, K. C.: Electron microscopy of herpes simplex virus. J. Virol. *2:* 1172–1184 (1968).

487 NII, S.; ROSENKRANZ, H. S.; MORGAN, C., and ROSE, H. M.: Electron microscopy of herpes simplex virus. III. Effect of hydroxyurea. J. Virol. *2:* 1163–1171 (1968).

488 NOBLE, J. and LONBERG-HOLM, K.: Interactions of components of human rhinovirus type 2 with HeLa cells. Virology *51:* 270–278 (1973).

489 NOMURA, S.: Interaction of respiratory syncytial virus with polycations: enhancement of infectivity with diethylaminoethyl dextran. Proc. Soc. exp. Biol. Med. *128:* 163–166 (1968).

490 NORRBY, E.: The relationship between the soluble antigens and the virion of adenovirus type 3. I. Morphological characteristics. Virology *28:* 236–248 (1966).

491 NORRBY, E.: The relationship between the soluble antigens and the virion of adenovirus type 3. II. Identification and characterization of an incomplete hemagglutinin. Virology *30:* 608–617 (1966).

492 Norrby, E.: Biological significance of structural adenovirus components. Curr. Top. Microbiol. Immunol. *43:* 1–43 (1968).

493 Norrby, E.: The structural and functional diversity of adenovirus capsid components. J. gen. Virol. *5:* 221–236 (1969).

494 Norrby, E.: The effect of carbobenzoxy tripeptide on the biological activities of measles virus. Virology *44:* 599–608 (1971).

495 Norrby, E. C. J.; Magnusson, P.; Falksveden, L. G., and Gronberg, M.: Separation of measles virus components by equilibrium centrifugation in CsCl gradients. II. Studies on the large and the small hemagglutinin. Arch. ges. Virusforsch. *14:* 462–473 (1964).

496 Norrby, E.; Marusyk, H., and Hammarskjöld, M.-L.: The relationship between the soluble antigens and the virion of adenovirus type 3. V. Identification of antigen specificities available at the surface of virions. Virology *38:* 477–482 (1969).

497 Norrby, E. and Skaaret, P.: The relationship between soluble antigens and the virion of adenovirus type 3. III. Immunological identification of fiber antigen and isolated vertex capsomer antigen. Virology *32:* 489–502 (1967).

498 Nozima, T. and Yasui, K.: Reaction with mannan of a live Japanese encephalitis virus vaccine. Acta virol., Praha *15:* 174 (1970).

499 Nozima, T.; Yasui, K., and Homma, R.: Reaction of several saccharides with Japanese encephalitis virus. Acta virol., Praha *12:* 296–300 (1968).

500 Ogston, A. G.: On uncertainties inherent in the determination of the efficiency of collision between virus particles and cells. Biochim. biophys. Acta *66:* 279–281 (1963).

501 Okada, Y. and Kim, J.: Interaction of concanavalin A with enveloped viruses and host cells. Virology *50:* 507–515 (1972).

502 Okada, Y. and Murayama, F.: Requirement of calcium ions for the cell fusion reaction of animal cells by HVJ. Exp. Cell Res. *44:* 527–551 (1966).

503 Okada, Y.; Murayama, F., and Yamada, K.: Requirement of energy for the cell fusion reaction of Ehrlich ascites tumor cells by HVJ. Virology *27:* 115–130 (1966).

504 Olshevsky, U. and Becker, Y.: Herpes simplex virus structural proteins. Virology *40:* 948–960 (1970).

505 Olshevsky, U. and Becker, Y.: Surface glycopeptides in the envelope of herpes simplex virions. Virology *50:* 277–279 (1972).

506 Onodera, T.; Hinuma, Y., and Ishida, N.: Ineffectiveness of antibody on the intracellular development of influenza virus W5 strain in Ehrlich ascites cells in mice. J. Immunol. *92:* 648–656 (1964).

507 Oram, J. D.; Ellwood, D. C.; Appleyard, G., and Stanley, J. L.: Agglutination of an arbovirus by concanavalin A. Nature new Biol. *233:* 50–51 (1971).

508 Osterman, J. V.; Wadell, A., and Aposhian, H. V.: DNA and gene therapy. Uncoating of polyoma pseudovirus in mouse embryo cells. Proc. nat. Acad. Sci., Wash. *67:* 37–40 (1970).

509 Oxford, J. S. and Schild, G. C.: Inhibition of the growth of influenza and rubella viruses by amines and ammonium salts. Brit. J. exp. Path. *48:* 235–243 (1967).

510 Ozaki, Y. and Kumagai, K.: Effect of polyions on Japanese encephalitis virus. Difference in interaction of virus with DEAE-dextran and dextran sulfate between PS [porcine kidney stable] cell-adapted and nonadapted virus. Arch. ges. Virusforsch. *39:* 83–91 (1972).

511 Ozanne, B. and Sambrook, J.: Binding of radioactively labelled concanavalin A and wheat germ agglutinin to normal and virus-transformed cells. Nature new Biol. *232:* 156–160 (1971).

512 Pagano, J. S. and Vaheri, A.: Enhancement of infectivity of poliovirus RNA with diethylaminoethyl-dextran (DEAE-D). Arch. ges. Virusforsch. *17:* 456–464 (1965).

513 Paranchych, W.; Krahn, P. M., and Bradley, R. D.: Stages in phage R17 infection. Virology *41:* 465–473 (1970).

514 Penman, S.; Becker, Y., and Darnell, J. E.: A cytoplasmic structure involved in the synthesis and assembly of poliovirus components. J. molec. Biol. *8:* 541–555 (1964).

515 Penman, S.; Scherrer, K.; Becker, Y., and Darnell, J. E.: Polyribosomes in normal and poliovirus-infected HeLa cells and their relationship to messenger RNA. Proc. nat. Acad. Sci., Wash. *49:* 654–662 (1963).

516 Penman, S. and Summers, D. F.: Effects on host cell metabolism following synchronous infection with poliovirus. Virology *27:* 614–620 (1965).

517 Pereira, H. G. and Figueiredo, M. V. T. de: Mechanism of hemagglutination by adenovirus types 1, 2, 4, 5, and 6. Virology *18:* 1–8 (1962).

518 Peries, J. R. and Chany, C.: Studies on measles viral hemagglutinin. Proc. Soc. exp. Biol. Med. *110:* 477–482 (1962).

519 Pernis, B.; Forni, L., and Amante, L.: Immunoglubulin spots on the surface of rabbit lymphocytes. J. exp. Med. *132:* 1001–1018 (1970).

520 Pettersson, U. and Höglund, S.: Structural proteins of adenoviruses. III. Purification and characterization of the adenovirus type 2 penton antigen. Virology *39:* 90–106 (1969).

521 Pettersson, U.; Philipson, L., and Höglund, S.: Structural proteins of adenoviruses. II. Purification and characterization of the adenovirus type 2 fiber antigen. Virology *35:* 204–215 (1968).

522 Philipson, L.: Studies on the mechanism of haemagglutination by ECHO viruses. I. Evidence for enzymatic degradation by chymotrypsin of the receptor for ECHO viruses. Arch. ges. Virusforsch. *9:* 251–260 (1959).

523 Philipson, L.: Early interaction of animal viruses and cells; in Melnick Progr. med. Virol., vol. 5, pp. 43–78 (Karger, Basel 1963).

524 Philipson, L.: Interaction between poliovirus and immunoglobulins. II. Basic aspects of virus-antibody interaction. Virology *28:* 35–46 (1966).

525 Philipson, L.: Attachment and eclipse of adenovirus. J. Virol. *1:* 868–875 (1967).

526 Philipson, L.: Virus inhibition by antibody; in Melnick Int. Virol., vol. 1, pp. 90–92 (Karger, Basel 1968).

527 Philipson, L.; Beatrice, S. T., and Crowell, R. L.: A structural model of picornaviruses as suggested from an analysis of urea-degraded virions and procapsids of coxsackievirus B3. Virology *54:* 69–79 (1973).

528 PHILIPSON, L. and BENGTSSON, S.: Interaction of enteroviruses with receptors from erythrocytes and host cells. Virology *18:* 457–469 (1962).

529 PHILIPSON, L.; BENGTSSON, S.; BRISHAMMAR, S.; SVENNERHOLM, L., and ZETTERQVIST, Ö.: Purification and chemical analysis of the erythrocyte receptor for hemagglutinating enteroviruses. Virology *22:* 580–590 (1964).

530 PHILIPSON, L. and CHOPPIN, P. W.: On the role of virus sulfhydryl groups in the attachment of enteroviruses to erythrocytes. J. exp. Med. *112:* 455–478 (1960).

531 PHILIPSON, L. and CHOPPIN, P. W.: Inactivation of enteroviruses by 2,3-dimercaptopropanol (BAL). Virology *16:* 405–413 (1962).

532 PHILIPSON, L. and LIND, M.: Enterovirus eclipse in a cell-free system. Virology *23:* 322–332 (1964).

533 PHILIPSON, L.; LONBERG-HOLM, K., and PETTERSSON, U.: Virus receptor interaction in an adenovirus system. J. Virol. *2:* 1064–1075 (1968).

534 PIRAINO, F.: The mechanism of genetic resistance of chick embryo cells to infection by Rous sarcoma virus – Bryan strain (BS-RSV). Virology *32:* 700–707 (1967).

535 PLAGEMANN, P. G. W. and SWIM, H. E.: Replication of mengovirus. I. Effects on synthesis of macromolecules by host cell. J. Bact. *91:* 2317–2326 (1966).

536 PLANTEROSE, D. N. and RYAN, K. O.: A 65S particle containing viral protein in cells infected with foot-and-mouth disease virus. Virology *26:* 372–374 (1965).

537 PODOPLEKIN, V. D.: Effect of tissue cultures on the hemagglutinating properties of ECHO viruses and mechanism of the interaction of these viruses with erythrocytes. Acta virol., Praha *8:* 254–262 (1964).

538 POGO, B. G. T. and DALES, S.: Two deoxyribonuclease activities within purified vaccinia virus. Proc. nat. Acad. Sci., Wash. *63:* 820–827 (1969).

538a POGO, B. G. T. and DALES, S.: Biogenesis of poxviruses: inactivation of host DNA polymerase by a component of the invading inoculum particle. Proc. Nat. Acad. Sci., Wash. *70:* 1726–1729 (1973).

539 POLLOCK, M. E.; TREADWELL, P. W., and KENNY, G. E.: Mammalian cell cultures contaminated with pleuropneumonia-like organisms. II. Effect of PPLO on cell morphology in established monolayer cultures. Exp. Cell Res. *31:* 321–328 (1963).

540 PONS, M. W.: Isolation of influenza virus ribonucleoprotein from infected cells. Demonstration of the presence of negative-stranded RNA in viral RNP. Virology *46:* 149–160 (1971).

541 PONS, M. W.; SCHULZE, I. T., and HIRST, G. K.: Isolation and characterization of the ribonucleoprotein of influenza virus. Virology *39:* 250–259 (1969).

542 POOLE, A. R.; HOWELL, J. I., and LUCY, J. A.: Lysolecithin and cell fusion. Nature, Lond. *227:* 810–814 (1970).

543 PRAGE, L.; PETTERSSON, U.; HÖGLUND, S.; LONBERG-HOLM, K., and PHILIPSON, L.: Structural proteins of adenoviruses. IV. Sequential degradation of the adenovirus type 2 virion. Virology *42:* 341–358 (1970).

544 PRAGE, L.; PETTERSSON, U., and PHILIPSON, L.: Internal basic proteins in adenovirus. Virology *36:* 508–511 (1968).

545 PRINCE, A. M. and GINSBERG, H. S.: Studies on the cytotoxic effect of Newcastle disease virus (NDV) on Ehrlich ascites tumor cells. I. Characteristics of the virus-cell interaction. J. Immunol. *79:* 94–106 (1957).

546 QUERSIN-THIRY, L.: Action of anticellular sera on virus infections. I. Influence on homologous tissue cultures infected with various viruses. J. Immunol. *81:* 253–260 (1958).

547 QUERSIN-THIRY, L.: Interaction between cellular extracts and animal viruses. I. Kinetic studies and some notes on the specificity of the interaction. Acta virol., Praha *5:* 141–152 (1961).

548 QUERSIN-THIRY, L. and NIHOUL, E.: Interaction between cellular extracts and animal viruses. II. Evidence for the presence of different inactivators corresponding to different viruses. Acta virol., Praha *5:* 283–293 (1961).

549 QUIGLEY, J. P.; RIFKIN, D. B., and COMPANS, R. W.: Isolation and characterization of ribonucleoprotein substructures from Rous sarcoma virus. Virology *50:* 65–75 (1972).

550 QUIGLEY, J. P.; RIFKIN, D. B., and REICH, E.: Phospholipid composition of Rous sarcoma virus, host cell membranes and other enveloped RNA viruses. Virology *46:* 106–116 (1971).

550a RABINOVITCH, M.: The dissociation of the attachment and ingestion phases of phagocytosis by macrophages. Exp. Cell Res. *46:* 19–28 (1967).

551 RADWAN, A. I. and BURGER, D.: The complement-requiring neutralization of equine arteritis virus by late antisera. Virology *51:* 71–77 (1973).

552 RAYBAUD, N.; JASMIN, C.; HUPPERT, J.; CHERMANN, J. C.; MATHÉ, G., and RAYNAUD, M.: Study of antivirus activity of mineral condensed polyanions. II. Effect on vesicular stomatitis virus. Rev. europ. Et. clin. Biol. *17:* 295–299 (1972).

553 RHIM, J. S. and HAMMON, W. M.: Quantitative studies of Japanese B encephalitis virus in hamster kidney cell cultures. Proc. Soc. exp. Biol. Med. *112:* 419–424 (1963).

554 RIFKIN, D. B. and COMPANS, R. W.: Identification of the spike proteins of Rous sarcoma virus. Virology *46:* 485–489 (1971).

555 RIGHTHAND, F.: *In vitro* interactions between virus and enhancer. Infect. Immun. *6:* 805–809 (1972).

556 RIGHTHAND, F. and KARZON, D. T.: Enhanced efficiency of echovirus 6 by a transferable factor in cultured human cells. Proc. Soc. exp. Biol. Med. *124:* 1248–1254 (1967).

557 RITZI, E. and LEVINE, A. J.: Deoxyribonucleic acid replication in simian virus 40-infected cells. III. Comparison of simian virus 40 lytic infection in three different monkey kidney cell lines. J. Virol. *5:* 686–692 (1970).

558 ROBB, J. A.: Penetration-uncoating of simian virus 40 (SV40). Control by both cellular and viral functions. J. Cell Biol. *47:* 172a (1970).

559 ROBB, J. A. and HUEBNER, K.: Effect of cell chromosome number on simian virus 40 replication. Exp. Cell Res. *81:* 120–126 (1973).

560 ROBB, J. A. and MARTIN, R. G.: Genetic analysis of simian virus 40. III. Characterization of a temperature-sensitive mutant blocked at an early stage of productive infection in monkey cells. J. Virol. *9:* 956–968 (1972).

561 ROBBINS, P. W. and MACPHERSON, I. A.: Glycolipid synthesis in normal and transformed animal cells. Proc. roy. Soc. Lond. B *177:* 49–58 (1971).

562 ROBINSON, W. S.: Ribonucleic acid polymerase activity in Sendai virions and nucleocapsid. J. Virol. *8:* 81–86 (1971).

563 ROBINSON, W. S.: Sendai virus RNA synthesis and nucleocapsid formation in the presence of cycloheximide. Virology *44:* 494–502 (1971).
564 ROESING, T. G. and CROWELL, R. L.: Elution and uncoating of coxsackievirus B3 by isolated HeLa cell plasma membranes. Proc. 73rd Ann. Meet. Amer. Soc. Microbiol. 1973, p. 212 (V104).
565 ROIZMAN, B.: The role of hormones in viral infections. II. Acceleration of viral adsorption and penetration into cells treated with thyroid hormone *in vitro.* Proc. nat. Acad. Sci., Wash. *48:* 973–977 (1962).
566 ROIZMAN, B.: Herpes viruses; in LEVY Biochemistry of viruses, pp. 415–482 (Marcel Dekker, New York 1969).
567 ROIZMAN, B. and ROANE, P. R., Jr.: The multiplication of herpes simplex virus. II. The relation between protein synthesis and the duplication of viral DNA in infected HEp-2 cells. Virology *22:* 262–269 (1964).
568 ROIZMAN, B. and SPEAR, P. G.: Herpesviruses: current information on the composition and structure; in MARAMOROSCH and KURSTAK Comparative virology, pp. 135–167 (Academic Press, New York 1971).
569 ROIZMAN, B.; SPRING, S. B., and SCHWARTZ, J.: The herpes virion and its precursors made in productively and in abortively infected cells. Fed. Proc. *28:* 1890–1898 (1969).
570 ROSEN, L.: Hemagglutination of adenoviruses. Virology 5: 574–577 (1958).
571 ROSEN, L.: A hemagglutination-inhibition technique for typing adenoviruses. Amer. J. Hyg. *71:* 120–128 (1960).
572 ROSEN, L.; HOVIS, J. F., and BELL, J. A.: Further observations on typing adenovirus and a description of two possible additional serotypes. Proc. Soc. exp. Biol. Med. *110:* 710–713 (1962).
573 ROSENBLITH, J. Z.; YIN, H. H.; UKENA, T. E.; BERLIN, R. D., and KARNOVSKY, M. J.: Topographical alteration of surface binding sites by concanavalin A in virally-transformed cells. Fed. Proc. *32:* Abstr. 3722 (1973).
573a ROTMAN, M. B. and CELADA, F.: Antibody-mediated activation of a defective $\beta$-D-galactosidase extracted from an *Escherichia coli* mutant. Proc. nat. Acad. Sci., Wash. *60:* 660–667 (1968).
574 ROTT, R. und MULLER, G.: Über den toxischen Effekt des Newcastle Disease Virus. Arch. ges. Virusforsch. *17:* 139–154 (1965).
575 ROWLANDS, D. J.; CARTWRIGHT, B., and BROWN, F.: Evidence for an internal antigen in foot-and-mouth disease virus. J. gen. Virol. *4:* 479–487 (1969).
576 ROWLANDS, D. J.; SANGER, D. V., and BROWN, F.: Relationship of the antigenic structure of foot-and-mouth disease virus to the process of infection. J. gen. Virol. *13:* 85–93 (1971).
577 RUBIN, H. and FRANKLIN, R. M.: On the mechanism of Newcastle disease virus neutralization by immune serum. Virology *3:* 84–95 (1957).
578 RUECKERT, R. R.: Picornaviral architecture; in MARAMOROSCH and KURSTAK Comparative virology, pp. 255–300 (Academic Press, New York 1971).
579 RUSSELL, W. C.; MCINTOSH, K., and SKEHEL, J. J.: The penetration and properties of adenovirus cores. J. gen. Virol. *11:* 35–46 (1971).
580 RUSSELL, W. C.; SKEHEL, J. J.; MACHADO, R., and PEREIRA, H. G.: Phosphorylated polypeptides in adenovirus-infected cells. Virology *50:* 931–934 (1972).

581 Russell, W.C.; Valentine, R. C., and Pereira, H. G.: The effect of heat on the anatomy of the adenovirus. J. gen. Virol. *1:* 509–522 (1967).

581a Sagik, B. P. and Levine, S.: The interaction of Newcastle disease virus (NDV) with chicken erythrocytes: attachment, elution, and hemolysis. Virology *3:* 401–416 (1957).

582 Sakiyama, H. and Burge, B. W.: Comparative studies of the carbohydrate-containing components of 3T3 and simian virus 40 transformed 3T3 mouse fibroblasts. Biochemistry *11:* 1366–1377 (1972).

583 Sakiyama, H.; Gross, S. K., and Robbins, P. W.: Glycolipid synthesis in normal and virus-transformed hamster cell lines. Proc. nat. Acad. Sci., Wash. *69:* 872–876 (1972).

584 Salzman, N. P.; Shatkin, A. J., and Sebring, E. D.: The synthesis of a DNA-like RNA in the cytoplasm of HeLa cells infected with vaccinia virus. J. molec. Biol. *8:* 405–416 (1964).

585 Sarkar, N. H.; Nowinski, R. C., and Moore, D. H.: Helical nucleocapsid structure of the oncogenic ribonucleic acid viruses (oncornaviruses). J. Virol. *8:* 564–572 (1971).

586 Sarov, I. and Joklik, W. K.: Studies on the nature and location of the capsid polypeptides of vaccinia virions. Virology *50:* 579–592 (1972).

587 Sarov, I. and Joklik, W. K.: Characterization of intermediates in the uncoating of vaccinia virus DNA. Virology *50:* 593–602 (1972).

588 Satoh, C.; Duff, R.; Rapp, F., and Davidson, E. A.: Production of mucopolysaccharides by normal and transformed cells. Proc. nat. Acad. Sci., Wash. *70:* 54–56 (1973).

589 Scandella, C. J.; Devaux, P., and McConnell, H. M.: Rapid lateral diffusion of phospholipids in rabbit sarcoplasmic reticulum. Proc. nat. Acad. Sci., Wash. *69:* 2056–2060 (1972).

589a Schaffer, F. L. and Hackett, A. J.: Early events in poliovirus-HeLa cell interaction: Acridine orange photosensitization and detergent extraction. Virology *21:* 124–126 (1963).

590 Scharff, M. D.; Maizel, J. V., and Levintow, L.: Physical and immunological properties of a soluble precursor of the poliovirus capsid. Proc. nat. Acad. Sci., Wash. *51:* 329–337 (1964).

591 Scheele, C. M. and Pfefferkorn, E. R.: Kinetics of incorporation of structural proteins into Sindbis virions. J. Virol. *3:* 365–375 (1969).

592 Scheid, A.; Caliguiri, L. A.; Compans, R. W., and Choppin, P. W.: Isolation of paramyxovirus glycoproteins. Association of both hemagglutinating and neuraminidase activities with the larger SV5 glycoprotein. Virology *50:* 640–652 (1972).

593 Scheid, A. and Choppin, P. W.: Isolation and purification of the envelope proteins of Newcastle disease virus. J. Virol. *11:* 263–271 (1973).

594 Schlesinger, M.: Adsorption of bacteriophages to homologous bacteria. II. Quantitative investigations of adsorption velocity and saturation. Estimation of the particle size of the bacteriophage. Z. Hyg. ImmunForsch. *114:* 149–160 (1932). English translation from Stent, G. S. (ed.): Papers on bacterial viruses, pp. 26–36 (Little, Brown, Boston 1960).

595 Schlesinger, M. J.; Schlesinger, S., and Burge, B. W.: Identification of a second glycoprotein in Sindbis virus. Virology *47:* 539–541 (1972).

596 Schlesinger, S. A. and Schlesinger, M. J.: Formation of Sindbis virus proteins: Identification of a precursor for one of the envelope proteins. J. Virol. *10:* 925–932 (1972).

597 Schmidt, N. J.; Dennis, J.; Hoffman, M. N., and Lennette, E. H.: Inhibitors of echovirus and reovirus hemagglutination. I. Inhibitors in tissue culture fluids. J. Immunol. *93:* 367–376 (1964).

598 Schmidt, N. J.; Dennis, J.; Hoffman, M. N., and Lennette, E. H.: Inhibitors of echovirus and reovirus hemagglutination. II. Serum and phospholipid inhibitors. J. Immunol. *93:* 377–386 (1964).

599 Scholtissek, C.; Drzeniek, R., and Rott, R.: Stepwise inactivation of an influenza virus by serial undiluted passages. Virology *30:* 313–314 (1966).

600 Scholtissek, C. and Rott, R.: Ribonucleic acid nucleotidyl transferase induced in chick fibroblasts after infection with an influenza virus. J. gen. Virol. *4:* 125–137 (1969).

601 Schulze, I. T.: Reversible inhibition of type 2 dengue virus by agar polysaccharide. Virology *22:* 79–90 (1964).

602 Schulze, I. T.: The structure of influenza virus. I. The polypeptides of the virion. Virology *42:* 890–904 (1970).

603 Schulze, I. T.: The structure of influenza virus. II. A model based on the morphology and composition of subviral particles. Virology *47:* 181–196 (1972).

604 Schwartz, J. and Roizman, B.: Concerning the egress of herpes simplex virus from infected cells. Electron and light microscope observations. Virology 38: 42–49 (1969).

605 Schwerdt, C. E. and Fogh, J.: The ratio of physical particles per infectious unit observed for poliomyelitis virus. Virology *4:* 41–52 (1957).

606 Shatkin, A. J.: Actinomycin D and vaccinia virus infection of HeLa cells. Nature, Lond. *199:* 357–358 (1963).

607 Shatkin, A. J. and LaFiandra, A. J.: Transcription by infectious subviral particles of reovirus. J. Virol. *10:* 698–706 (1972).

608 Shatkin, A. J. and Sipe, J. D.: RNA polymerase activity in purified reoviruses. Proc. nat. Acad. Sci., Wash. *61:* 1462–1469 (1968).

609 Shen, L. and Ginsburg, V.: Release of sugars from HeLa cells by trypsin; in Manson Biological properties of the mammalian surface membrane, pp. 67–71 (Wistar Institute Press, Philadelphia 1968).

610 Shigematsu, T. and Dmochowski, L.: Studies on the acid mucopolysaccharide coat of viruses and transformed cells. Cancer, Philad. *31:* 165–174 (1973).

611 Shimono, H. and Kaplan, A. S.: Correlation between the synthesis of DNA and histones in polyoma virus-infected mouse embryo cells. Virology *37:* 690–694 (1969).

612 Shirman, G. A.; Maslova, S. V.; Gavrilovskaya, I. N., and Agol, V. I.: Stimulation of restricted reproduction of EMC virus in HeLa cells by non-replicating poliovirus. Virology *51:* 1–10 (1973).

613 Shirobokov, V. P.: Differentiation of Coxsackieviruses based on the character of adsorption onto bentonite. Acta virol., Praha *12:* 185 (1968).

614 Shortridge, K. F. and Biddle, F.: A human serum inhibitor of adenovirus hemagglutination. Arch. ges. Virusforsch. *25:* 148–159 (1968).

615 Siegert, R. und Falke, D.: Elektronenmikroskopische Untersuchungen über die Entwicklung des Herpesvirus hominis in Kulturzellen. Arch. ges. Virusforsch. *19:* 230–249 (1966).

616 Silverstein, S. C.; Astell, C.; Levin, D. H.; Schonberg, M., and Acs, G.: The mechanism of reovirus uncoating and gene activation *in vivo*. Virology *47:* 797–806 (1972).

617 Silverstein, S. C. and Castle, D.: The effect of antibody on the phagocytosis and processing of vaccinia virus by mouse peritoneal macrophages. J. Cell Biol. *47:* 192a (1970).

618 Silverstein, S. C. and Dales, S.: The penetration of reovirus RNA and initiation of its genetic function in L-strain fibroblasts. J. Cell Biol. *36:* 197–230 (1968).

619 Silverstein, S. C. and Marcus, P. I.: Early stages of Newcastle disease virus-HeLa cell interaction. An electron microscopic study. Virology *23:* 370–380 (1964).

620 Silverstein, S. C.; Schonberg, M.; Levin, D. H., and Acs, G.: The reovirus replicative cycle. Conservation of parental RNA and protein. Proc. nat. Acad. Sci., Wash. *67:* 275–281 (1970).

621 Simons, K. and Kääriäinen, L.: Characterization of the Semliki Forest virus core and envelope proteins. Biochem. biophys. Res. Commun. *38:* 981–988 (1970).

622 Simpson, R. W.; Hauser, R. E., and Dales, S.: Viropexis of vesicular stomatitis virus by L cells. Virology *37:* 285–290 (1969).

623 Singer, S. J.: The molecular organization of biological membranes; in Rothfield Structure and function of biological membranes, pp. 145–222 (Academic Press, New York 1971).

624 Singer, S. J.: Yes, cell membranes are fluid. Science *180:* 983 (1973).

625 Singer, S. J. and Nicolson, G. L.: The fluid mosaic model of the structure of cell membranes; in Day and Good Membranes and viruses in immunopathology, pp. 7–47 (Academic Press, New York 1972).

626 Singer, S. J. and Nicolson, G. L.: The fluid mosaic model of the structure of cell membranes. Science *175:* 720–731 (1972).

627 Skehel, J. J.: The characterization of subviral particles derived from influenza virus. Virology *44:* 409–417 (1971).

628 Skehel, J. and Schild, G. C.: The polypeptide composition of influenza A viruses. Virology *44:* 396–408 (1971).

629 Soloviev, V. D.; Krispin, T. I.; Zaslavsky, V. G., and Agol, V. I.: Mechanism of resistance to enteroviruses of some primate cells in tissue culture. J. Virol. *2:* 553–557 (1968).

630 Speth, V.; Wallach, D. F. H.; Weidekamm, E., and Knüfermann, H.: Micromorphologic consequence following perturbation of erythrocyte membranes by trypsin, phospholipase A, lysolecithin, sodium dodecyl sulfate and saponin. A correlated freeze-etching and biochemical study. Biochim. biophys. Acta *255:* 386–394 (1972).

631 Spring, S. B. and Roizman, B.: Herpes simplex virus products in productive and abortive infection. III. Differentiation of infectious virus derived from nucleus and cytoplasm with respect to stability and size. J. Virol. *2:* 979–985 (1968).

632 SPRING, S. B.; ROIZMAN, B., and SCHWARTZ, J.: Herpes simplex virus products in productive and abortive infection. II. Electron microscopic and immunological evidence for failure of virus envelopment as a cause of abortive infection. J. Virol. *2:* 384–392 (1968).

633 SPRINGER, G. F.: Enzymatic and nonenzymatic alterations of erythrocyte surface antigens. Bact. Rev. *27:* 191–227 (1963).

634 SPRINGER, G. F.; HUPRIKAR, S. V., and TEGTMEYER, H.: The biochemical and genetic basis of human blood group MN specificities and their relation to infectious mononucleosis- and oncogenic virus-receptors; in JAMIESON and GREENWALT Glycoproteins of blood cells and plasma, pp. 35–49 (Lippincott, Philadelphia 1971).

635 SPRINGER, G. F.; NAGAI, Y., and TEGTMEYER, H.: Isolation and properties of human blood-group NN and meconium-Vg antigens. Biochemistry *5:* 3254–3272 (1966).

636 SPRINGER, G. F.; SCHWICK, H. G., and FLETCHER, M. A.: The relationship of the influenza virus inhibitory activity of glycoproteins to their molecular size and sialic acid content. Proc. nat. Acad. Sci., Wash. *64:* 634–641 (1969).

637 STANLEY, P.; GANDHI, S. S., and WHITE, D. O.: The polypeptides of influenza virus. VII. Synthesis of the hemagglutinin. Virology *53:* 92–106 (1973).

638 STANLEY, P. and HASLAM, E. A.: The polypeptides of influenza virus. V. Localization of polypeptides in the virion by iodination techniques. Virology *46:* 764–773 (1971).

639 STASNY, J. T.; NEURATH, A. R., and RUBIN, B. A.: Effect of formamide on the capsid morphology of adenovirus types 4 and 7. J. Virol. *2:* 1429–1442 (1968).

640 STECK, T. L.; FAIRBANKS, G., and WALLACH, D. F. H.: Disposition of the major proteins in the isolated erythrocyte membrane. Proteolytic dissection. Biochemistry *10:* 2617–2624 (1971).

641 STECK, T. L. and FOX, C. F.: Membrane proteins; in FOX and KEITH Membrane molecular biology, pp. 27–75 (Sinauer Associates, Stamford 1972).

642 STECK, F. T. and RUBIN, H.: The mechanism of interference between an avian leukosis virus and Rous sarcoma virus. II. Early steps of infection by RSV of cells under conditions of interference. Virology *29:* 642–653 (1966).

643 STINSKI, M. F. and CUNNINGHAM, C. H.: Neutralizing antibody complex of infectious bronchitis virus. J. Immunol. *102:* 720–727 (1969).

643a STOFFEL, W.: Sphingolipids. Annu. Rev. Biochem. *40:* 57–82 (1971).

644 STOTT, E. J. and KILLINGTON, R. A.: Haemagglutination by rhinoviruses. Lancet *i:* 1369–1370 (1972).

645 STONE, J. D. and ADA, G. L.: Electrophoretic studies of virus-red cell interaction. Relationship between agglutinability and electrophoretic mobility. Brit. J. exp. Path. *33:* 428 (1952).

646 STONE, H. O. and KINGSBURY, D. W.: Stimulation of Sendai virion transcriptase by polyanions. J. Virol. *11:* 243–249 (1973).

647 STONE, H. O.; PORTNER, A., and KINGSBURY, D. W.: Ribonucleic acid transcriptases in Sendai virions and infected cells. J. Virol. *8:* 174–180 (1971).

648 STRAUSS, J. H.; BURGE, B. W.; PFEFFERKORN, E. R., and DARNELL, J. E.: Identification of the membrane protein and 'core' protein of Sindbis virus. Proc. nat. Acad. Sci., Wash. *59:* 533–540 (1968).
649 STURMAN, L. S. and TAMM, I.: Formation of viral ribonucleic acid and virus in cells that are permissive or nonpermissive for murine encephalomyelitis virus (GDVII). J. Virol. *3:* 8–16 (1969).
650 SUGIYAMA, T.; KORANT, B. D., and LONBERG-HOLM, K. K.: RNA virus gene expression and its control. Annu. Rev. Microbiol. *26:* 467–502 (1972).
651 SUMMERS, D. F.; MAIZEL, J. V., jr., and DARNELL, J. E., jr.: Evidence for virus-specific noncapsid proteins in poliovirus-infected HeLa cells. Proc. nat. Acad. Sci., Wash. *54:* 505–513 (1965).
652 SUMMERS, D. F.; MCELVAIN, N. F.; THORÉN, M. M., and LEVINTOW, L.: Incorporation of amino acids into polyribosome-associated protein in cytoplasmic extracts of poliovirus-infected HeLa cells. Biochem. biophys. Res. Commun. *15:* 290–295 (1964).
653 SUMMERS, M. D.: Apparent *in vivo* pathway of granulosis virus invasion and infection. J. Virol. *4:* 188–190 (1969).
654 SUMMERS, M. D.: Electron microscopic observations on granulosis virus entry, uncoating and replication processes during infection of the midgut cells of *Trichoplusia ni*. J. Ultrastruct. Res. *35:* 606–625 (1971).
655 SUNDQUIST, B.; PETTERSSON, U.; THELANDER, L., and PHILIPSON, L.: Structural proteins of adenovirus. IX. Molecular weight and subunit composition of adenovirus type 2 fiber. Virology *51:* 252–256 (1973).
656 SUSSENBACH, J. S.: Early events in the infection process of adenovirus type 5 in HeLa cells. Virology *33:* 567–574 (1967).
657 SUTTAJIT, M. and WINZLER, R. J.: Effect of modification of N-acetylneuraminic acid on the binding of glycoproteins to influenza virus and on susceptibility to cleavage by neuraminidase. J. biol. Chem. *246:* 3398–3404 (1971).
658 SVEHAG, S. E.: Reactivation of neutralized virus by fluorocarbon. Mechanism of action and demonstration of reduced reactivability with time of virus-antibody interaction. Virology *21:* 174–182 (1963).
659 SYDISKIS, R. J. and ROIZMAN, B.: Polysomes and protein synthesis in cells infected with a DNA virus. Science *153:* 76–78 (1966).
660 SYDISKIS, R. J. and ROIZMAN, B.: The disaggregation of host polyribosomes in productive and abortive infection with herpes simplex virus. Virology *32:* 678–686 (1967).
661 SZÁNTÓ, J.: Haemagglutination reaction with avian myeloblastosis virus. Acta virol., Praha *15:* 245–248 (1971).
662 SZÁNTÓ, J.; LAUČÍKOVÁ, O.; KOČIŠKOVÁ, D., and RÁTHOVÁ, V.: The effect of anti-inhibitor serum on the multiplication of influenza virus. Virology *10:* 149 (1960).
663 SZILÁGYI, J. F. and URYVAYEV, L.: Isolation of an infectious ribonucleoprotein from vesicular stomatitis virus containing an active RNA transcriptase. J. Virol. *11:* 279–286 (1973).
664 TAKEMOTO, K. K. and FABISCH, P.: Influence of acid polysaccharides on plaque formation by influenza A2 and B viruses. Proc. Soc. exp. Biol. Med. *114:* 811–814 (1963).

665 Takemoto, K. K. and Fabisch, P.: Inhibition of herpes virus by natural and synthetic acid polysaccharides. Proc. Soc. exp. Biol. Med. *116:* 140–144 (1964).

666 Takemoto, K. K. and Spicer, S. S.: Effects of natural and synthetic sulfated polysaccharides on viruses and cells. Ann. N. Y. Acad. Sci. *130:* 365–373 (1965).

667 Talbot, P. and Brown, F.: A model for foot-and-mouth disease virus. J. gen. Virol. *15:* 163–170 (1972).

668 Talbot, P.; Rowlands, D. J.; Burroughs, J. N.; Sangar, D. V., and Brown, F.: Evidence for a group protein in foot-and-mouth disease virus particles. J. gen. Virol. *19:* 369–380 (1973).

669 Tamm, I. and Caliguiri, L. A.: Mode of action of antiviral substances; in Sanders and Schaeffer Viruses affecting man and animals, pp. 136–152 (Green, St. Louis 1971).

670 Tanada, Y. and Luetenegger, R.: Multiplication of a granulosis virus in larval midgut cells of *Trichoplusia ni* and possible pathways of invasion into the hemocoel. J. Ultrastruct. Res. *30:* 589–600 (1970).

671 Taylor, J. M.; Hampson, A. W.; Layton, J. E., and White, D. O.: The polypeptides of influenza virus. IV. An analysis of nuclear accumulation. Virology *42:* 744–752 (1970).

672 Taylor, R. B.; Duffus, W. P. H.; Raff, M. C., and de Petris, S.: Redistribution and pinocytosis of lymphocyte surface immunoglobulin molecules induced by anti-immunoglobulin antibody. Nature new Biol. *233:* 225–229 (1971).

673 Tegtmeyer, P. and Enders, J. F.: Feline herpesvirus infection in fused cultures of naturally resistant human cells. J. Virol. *3:* 469–476 (1969).

674 Temin, H. M. and Baltimore, D.: RNA-directed DNA synthesis and RNA tumor viruses. Adv. Virus Res. *17:* 129–186 (1972).

675 Tennant, J. R. and Snell, G. D.: The H-2 locus and viral leukemogenesis as studied in oncogenic strains of mice. J. nat. Cancer Inst. *41:* 597–604 (1968).

676 Thomas, D. B.: Cyclic expression of blood group determinants in murine cells and their relationship to growth control. Nature, Lond. *233:* 317–321 (1971).

677 Thorne, H. V.: Electrophoretic study of the interaction of radioactive poliovirus with components of cultured cells. J. Bact. *185:* 1247–1255 (1963).

678 Thorne, H. V.: Cyclic variation in susceptibility of Balb-C 3T3 cells to polyoma virus. J. gen. Virol. *18:* 163–169 (1973).

679 Tiffany, J. M. and Blough, H. A.: Estimation of the number of surface projections on myxo- and paramyxoviruses. Virology *41:* 392–394 (1970).

680 Tiffany, J. M. and Blough, H. A.: Attachment of myxoviruses to artificial membranes. Electron microscopic studies. Virology *44:* 18–28 (1971).

681 Tillotson, J. R. and Lerner, A. M.: Effect of periodate oxidation on hemagglutinating and antibody-producing capacities of certain enteroviruses and reoviruses. Proc. nat. Acad. Sci., Wash. *56:* 1143–1150 (1966).

682 Timbury, M. C.: The growth of echovirus 1 in HEp-2 cells treated with antibody to the host cells. J. gen. Virol. *4:* 531–539 (1969).

683 Tischer, I.: Untersuchungen über die pH-Abhängigkeit der Haemagglutination und die Beziehung zwischen Oberflächenladung und Receptoraffinität einiger Myxoviren. Arch. ges. Virusforsch. *12:* 89–107 (1963).

684 TISCHER, I.: Erhöhte Empfindlichkeit von neuraminidasebehandelten Affenerythrocyten für den Nachweis der Haemagglutinationsaktivität von Masernvirus und einigen Adenoviren. Z. Bakt. Parasitkde I. Abt. *204:* 466–476 (1967).

685 TODARO, G. J.; LAZAR, G. K., and GREEN, H.: The initiation of cell division in a contact-inhibited mammalian cell line. J. cell. comp. Physiol. *66:* 325–334 (1965).

686 UCHIDA, S.; WATANABE, S., and KATO, M.: Incomplete growth of simian virus 40 in African green monkey kidney culture induced by serial undiluted passages. Virology *28:* 135–141 (1966).

687 VAHERI, A.; SEDWICK, W. D., and PLOTKIN, S. A.: Growth of rubella virus in BHK21 cells. II. Enhancing effect of DEAE-dextran, semicarbazide and low doses of metabolic inhibitors. Proc. Soc. exp. Biol. Med. *125:* 1092–1098 (1967).

688 VALENTINE, R. C. and ALLISON, A. C.: Virus particle adsorption. I. Theory of adsorption and experiments on the attachment of particles to non-biological surfaces. Biochim. biophys. Acta *34:* 10–23 (1959).

689 VALENTINE, R. C. and PEREIRA, H. G.: Antigens and structure of the adenovirus. J. molec. Biol. *13:* 13–20 (1965).

690 VAN BRUGGEN, J. T.: Chemistry of the membrane; in BROWN Chemistry of the cell interface, pp. 1–32 (Academic Press, New York 1971).

691 VAN DER NOORDAA, J.; ENDERS, J. F., and DIAMANDOPOULOS, G. T.: Increased resistance to herpes simplex virus of hamster and human cells transformed by SV40. Proc. Soc. exp. Biol. Med. *122:* 915–920 (1966).

692 VAUGHAN, M. H., jr.; SOEIRO, R.; WARNER, J. R., and DARNELL, J. E., jr.: The effects of methionine deprivation on ribosome synthesis in HeLa cells. Proc. nat. Acad. Sci., Wash. *58:* 1527–1534 (1967).

693 VERLINDE, J. D. et DE BAAN, P.: Sur l'hémagglutination par des virus poliomyélitiques murins et la destruction enzymatique des récepteurs de virus poliomyélitique de la cellule réceptive. Ann. Inst. Pasteur *77:* 632–641 (1949).

694 VOGT, P. K.: DEAE-dextran. Enhancement of cellular transformation induced by avian sarcoma viruses. Virology *33:* 175–177 (1967).

695 VOGT, M. and DULBECCO, R.: Properties of a HeLa cell culture with increased resistance to poliomyelitis virus. Virology *5:* 425–434 (1958).

696 VON MAGNUS, P.: Incomplete forms of influenza virus. Adv. Virus Res. *2:* 59–79 (1954).

697 WACKER, W. B.; LIPTON, M. M., and STEIGMAN, A. J.: Demonstration of a toxin in mumps virus and its dissociation from infective and hemagglutinating particles. J. Immunol. *89:* 525–530 (1962).

698 WADELL, G.: Hemagglutination with adenovirus serotypes belonging to Rosen's subgroups II and III. Proc. Soc. exp. Biol. Med. *132:* 413–421 (1969).

699 WADELL, G.: Structural and biological properties of capsid components of human adenoviruses; Thesis, Stockholm (1970).

700 WADELL, G. and NORRBY, E.: The soluble hemagglutinins of adenoviruses belonging to Rosen's subgroup III. I. The rapidly sedimenting hemagglutinin. Arch. ges. Virusforsch. *26:* 53–62 (1969).

701 WADELL, G.; NORRBY, E., and SKAARET, P.: The soluble hemagglutinins of adenoviruses belonging to Rosen's subgroup III. I. The rapidly sedimenting hemagglutinin. Arch. ges. Virusforsch. *26:* 32–52 (1969).

702 WAGNER, E. K. and ROIZMAN, B.: Ribonucleic acid synthesis in cells infected with herpes simplex virus. I. Patterns of ribonucleic acid synthesis in productively infected cells. J. Virol. *4:* 36–46 (1969).

703 WAGNER, R.; ROSENBERG, M., and ESTENSEN, R.: Endocytosis in Chang liver cells. Quantitation by sucrose-$^{3}$H uptake and inhibition by cytochalasin B. J. Cell Biol. *50:* 804–817 (1971).

704 WAGNER, R. R. and HUANG, A. S.: Inhibition of RNA and interferon synthesis in Krebs-2 cells infected with vesicular stomatitis virus. Virology *28:* 1–10 (1966).

705 WAGNER, R. R.; SCHNAITMAN, T. A., and SNYDER, R. M.: Structural proteins of vesicular stomatitis viruses. J. Virol. *3:* 395–403 (1969).

706 WALL, R. and TAYLOR, M. W.: Host-dependent restriction of mengovirus replication. J. Virol. *4:* 681–687 (1969).

707 WALL, R. and TAYLOR, M. W.: Mengovirus RNA synthesis in productive and restrictive cell lines. Virology *42:* 78–86 (1970).

708 WALLACH, D. F. H.: The dispositions of proteins in the plasma membranes of animal cells. Analytical approaches using controlled peptidolysis and protein labels. Biochim. biophys. Acta *265:* 61–83 (1972).

709 WALLBANK, A. M.: $\alpha$-Methyl-l-adamantane-methylamine hydrochloride. Inhibition of Rous and Esh sarcoma viruses in cell culture. Proc. Soc. exp. Biol. Med. *131:* 1025–1027 (1969).

710 WALLIS, C. and MELNICK, J. L.: Herpesvirus neutralization. The role of complement. J. Immunol. *107:* 1235–1242 (1971).

711 WALLIS, C.; SMITH, K. O., and MELNICK, J. L.: Reovirus activation by heating and inactivation by cooling in $MgCl_2$ solutions. Virology *22:* 608–619 (1964).

712 WANG, R.; POLLACK, R.; KUSANO, T., and GREEN, H.: Human-mouse hybrid cell lines and susceptibility to poliovirus. I. Conversion from polio sensitivity to polio resistance accompanying loss of human gene-dependent polio receptors. J. Virol. *5:* 677–681 (1970).

713 WARREN, L.; FUHRER, J. P., and BUCK, C. A.: Surface glycoproteins of normal and transformed cells. A difference determined by sialic acid and a growth-dependent sialyl transferase. Proc. nat. Acad. Sci., Wash. *69:* 1838–1842 (1972).

714 WATERSON, A. P.: in Seminar on the epidemiology and prevention of measles and rubella. Arch. ges. Virusforsch. *16:* 57–80 (1965).

715 WATSON, D. H.: The separation of herpesvirus-specific antigens by polyacrylamide gel electrophoresis. J. gen. Virol. *4:* 151–161 (1969).

716 WATSON, D. H.; RUSSELL, W. C., and WILDY, P.: Electron microscopic particle counts on herpes virus using the phosphotungstate negative staining technique. Virology *19:* 250–260 (1963).

717 WAY, H. J.: Enhancement of encephalomyocarditis plaques by saponin. J. gen. Virol. *5:* 557–559 (1969).

718 WEBSTER, R. G.: The structure and function of the haemagglutinin and neuraminidase of influenza virus; in BARRY and MAHY The biology of large RNA viruses, pp. 53–74 (Academic Press, New York 1970).

719 WEBSTER, R. G.: Estimation of the molecular weights of the polypeptide chains from the isolated hemagglutinin and neuraminidase subunits of influenza viruses. Virology *40:* 643–654 (1970).

720 WEBSTER, R. G. and DARLINGTON, R. W.: Disruption of myxoviruses with Tween 20 and isolation of biologically active hemagglutinin and neuraminidase subunits. J. Virol. *4:* 182–187 (1969).

721 WEIL, R.; MICHEL, M. R., and RUSCHMANN, G. K.: Induction of cellular DNA synthesis by polyoma virus. Proc. nat. Acad. Sci., Wash. *53:* 1468–1475 (1965).

722 WEIL, R.; PÉTURSSON, G.; KARA, J., and DIGGELMANN, H.: The interaction of polyoma virus with the genetic apparatus of host cells; in COLTER and PARANCHYCH The molecular biology of viruses, pp. 593–626 (Academic Press, New York 1967).

723 WERTZ, G. W. and YOUNGNER, J. S.: Inhibition of protein synthesis in L cells infected with vesicular stomatitis virus. J. Virol. *9:* 85–89 (1972).

724 WESTWOOD, J. C. N.; HARRIS, W. J.; ZWARTOUW, H. T.; TITMUSS, D. H. J., and APPLEYARD, G.: Studies on the structure of vaccinia virus. J. gen. Microbiol. *34:* 67–78 (1964).

725 WHEELOCK, E. F. and TAMM, I.: Biochemical basis for alterations in structure and function of HeLa cells infected with Newcastle disease virus. J. exp. Med. *114:* 617–632 (1961).

726 WHITE, D. O.; SCHARFF, M. D., and MAIZEL, J. V., jr.: The polypeptides of adenoviruses. III. Synthesis in infected cells. Virology *38:* 395–406 (1969).

727 WHITEHEAD, P. H.; FLEWETT, T. H.; FOSTER, J. R., and SAMMONS, H. G.: Inhibition of influenza virus haemagglutination by polymerized orosomucoid. Nature, Lond. *208:* 915–916 (1965).

728 WHITEHEAD, P. H. and WINZLER, R. J.: Inhibition of viral haemagglutination by aggregated orosomucoid. Arch. Biochem. Biophys. *126:* 657–663 (1968).

729 WILLEMS, F. T. C.; MELNICK, J. L., and RAWLS, W. E.: Replication of poliovirus in phytohemagglutinin-stimulated human lymphocytes. J. Virol. *3:* 451–457 (1969).

730 WILLEMS, M. and PENMAN, S.: The mechanism of host cell protein synthesis inhibition by poliovirus. Virology *30:* 355–367 (1966).

731 WILLEMS, F. T. C. and RAWLS, W. E.: Poliovirus inhibition of the phytohemagglutinin response in human lymphocytes. Arch. ges. Virusforsch. *27:* 352–363 (1969).

732 WILLS, E. J.; DAVIES, P.; ALLISON, A. C., and HASWELL, A. D.: Cytochalasin B fails to inhibit pinocytosis by macrophages. Nature new Biol. *240:* 58–60 (1972).

733 WILSON, D. E.: Inhibition of host-cell protein and ribonucleic acid synthesis by Newcastle disease virus. J. Virol. *2:* 1–6 (1968).

734 WILSON, J. N. and COOPER, P. D.: Aspects of the growth of poliovirus as revealed by the photodynamic effects of neutral red and acridine orange. Virology *21:* 135–145 (1963).

735 WINOCOUR, E. and ROBBINS, E.: Histone synthesis in polyoma and SV40-infected cells. Virology *40:* 307–315 (1970).

736 WINTERBURN, P. J. and PHELPS, C. F.: The significance of glycosylated proteins. Nature, Lond. *236:* 147–151 (1972).

737 WINZLER, R. J.: A glycoprotein in human erythrocyte membranes; in JAMIESON and GREENWALT Red cell membrane, structure and function, pp. 157–171 (Lippincott, Philadelphia 1969).

738 WINZLER, R. J.: The association of a glycoprotein component of erythrocyte stroma with the cell membrane; in SMITH and GOOD Cellular recognition, pp. 11–18 (Appleton Century Crofts, New York 1969).

739 WOODSON, B.: Recent progress in poxvirus research. Bact. Rev. *32:* 127–137 (1968).

740 WRIGLEY, N. G.; SKEHEL, J. J.; CHARLWOOD, P. A., and BRAND, C. M.: The size and shape of influenza virus neuraminidase. Virology *51:* 525–529 (1973).

741 YAHARA, I. and EDELMAN, G. M.: Restriction of the mobility of lymphocyte immunoglobulin receptors by concanavalin A. Proc. nat. Acad. Sci., Wash. *69:* 608–612 (1972).

742 YAMAZAKI, S. and WAGNER, R. R.: Action of interferon. Kinetics and differential effects on viral functions. J. Virol. *6:* 421–429 (1970).

743 YAOI, Y. and OGATA, M.: The fate of vesicular stomatitis virus in cultured chick embryo cells. J. gen. Virol. *16:* 419–422 (1972).

744 YASUI, K. and NOZIMA, T.: Partial *in vitro* reconstruction of Japanese encephalitis virus receptor. Acta virol., Praha *15:* 113–118 (1971).

745 YASUI, K.; NOZIMA, T.; HOMMA, R., and UEDA, S.: Effects of a mannosidase on the active site of Japanese encephalitis viral receptor. Acta virol., Praha *13:* 158 (1968).

746 YASUI, K.; NOZIMA, T.; HOMMA, R., and UEDA, S.: Extraction of active fragment from Japanese encephalitis viral receptor of susceptible cells. Acta virol., Praha *15:* 7–18 (1971).

747 YOUNGNER, J. S.: Adsorption of enteroviruses to water-insoluble basic lipids. Fed. Proc. *21:* 458 (1962).

748 ZAJAC, I. and CROWELL, R. L.: Effect of enzymes on the interaction of enteroviruses with living HeLa cells. J. Bact. *89:* 574–582 (1965).

749 ZAJAC, I. and CROWELL, R. L.: Location and regeneration of enterovirus receptors of HeLa cells. J. Bact. *89:* 1097–1100 (1965).

750 ZAJAC, I. and CROWELL, R. L.: Differential inhibition of attachment and eclipse activities of HeLa cells for enteroviruses. J. Virol. *3:* 422–428 (1969).

751 ZÁVADA, J.: Pseudotypes of vesicular stomatitis virus with the coat of murine leukaemia and of avian myeloblastosis viruses. J. gen. Virol. *15:* 183–191 (1972).

752 ZÁVADA, J.; ZÁVADA, Z.; MALIŘ, A., and KOĆENT, A.: VSV pseudotype produced in cell line derived from human mammary carcinoma. Nature new Biol. *240:* 124–125 (1972).

753 ZEE, Y. C. and TALENS, L.: Entry of infectious bovine rhinotracheitis virus into cells. J. gen. Virol. *11:* 59–63 (1971).

754 ZHDANOV, V. M. and BUKRINSKAYA, A. G.: Studies on the initial stage of virus-cell interaction. Acta virol., Praha *6:* 105–113 (1962).

755 ZHDANOV, V. M.; BUKRINSKAYA, A. G., and RAMENSKAYA, G. P.: Autoradiographic study of the penetration of Sendai virus into the cell. III. Use of virus preparations labeled with uracil-$^{14}C$, $^{32}P$, methionine-$^{35}S$ or cysteine-$^{35}S$. Acta virol., Praha *7:* 1–6 (1963).

756 ZIMMERMAN, E. F.; HEETER, M., and DARNELL, J. E.: RNA synthesis in poliovirus-infected cells. Virology *19:* 400–408 (1963).

# Cumulative Author Index